P9-DXI-167

SOME WE LOVE,

SOME WE HATE,

SOME WE EAT

SOME WE **LOVE**,
SOME WE **HATE**,
SOME WE **EAT**

Why It's So Hard to Think
Straight About Animals

Hal Herzog

HARPER

An Imprint of HarperCollins*Publishers*
www.harpercollins.com

4355 9987 9/10

FIRST EDITION

Designed by Cassandra J. Pappas

Library of Congress Cataloging-in-Publication data is available upon request.

ISBN 978-0-06-173086-3

10 11 12 13 14 ov/rrd 10 9 8 7 6 5 4 3 2 1

To Adam, Betsy, Katie, and most of all, Mary Jean,
to whom I owe everything

CONTENTS

Why Is It So Hard to Think Straight About Animals?

I like pondering our relationships with animals because they tell a lot about who we are.

                                                    —MARC BEKOFF

The way we think about other species often defies logic. Consider Judith Black. When she was twelve, Judith decided that it was wrong to kill animals just because they taste good. But what exactly is an animal? While it is obvious that dogs and cats and cows and pigs are animals, it was equally clear to Judith that fish were not. They just didn't *feel* like animals to her. So for the next fifteen years, this intuitive biological classification system enabled Judith, who has a PhD in anthropology, to think of herself as a vegetarian, yet still experience the joys of smoked Copper River salmon and lemon-grilled swordfish.

This twisted moral taxonomy worked fine until Judith ran into Joseph Weldon, a graduate student in biology. When they first met, Joseph, himself a meat eater, tried to convince Judith that there is not a shred of moral difference between eating a Cornish hen and a Chilean sea bass. After all, he reasoned, both birds and fish are vertebrates, have brains, and lead social lives. Despite his best efforts, he failed to convince her that from a culinary ethics perspective, a cod is a chicken is a cow.

Fortunately, their disagreement over the moral status of mahi mahi did not prevent them from falling in love. They married, and her new husband kept the fish-versus-fowl discussion going over the dinner table. After three years of philosophical to-and-fro, Judith sighed one evening and gave in: "OK, I see your point. Fish are animals."

But now she faced a difficult decision: She could either quit eating fish, or stop thinking of herself as a vegetarian. Something had to give. A week later, friends invited Joseph to a grouse hunt. Though he had no experience with a shotgun, he somehow managed to hit a bird on the fly, and, in grand caveman tradition, showed up at home, dead carcass in hand. Joseph then proceeded to pluck and cook the grouse, which he proudly served to his wife for dinner along with wild rice and a lovely raspberry sauce.

In an instant, fifteen years of moral high ground went down the drain. ("I am a sucker for raspberries," Judith told me.) The taste of roasted grouse opened the floodgates and there was no going back. Within a week, she was chowing down on cheeseburgers. Judith had joined the ranks of ex-vegetarians, a club that outnumbers current vegetarians in the United States by a ratio of three to one.

Then there is Jim Thompson, a twenty-five-year-old doctoral student in mathematics who was working on his dissertation when I met him. Before beginning graduate school, Jim had worked in a poultry research laboratory in Lexington, Kentucky, where one of his jobs consisted of dispatching baby chicks at the end of the experiments. For a while, this posed no problem for Jim. However, things changed one day when he was looking for a magazine to read on a plane and his mother handed him a copy of *The Animals' Agenda*, a magazine that advocated animal rights. He never ate meat again.

That was just the start. Over the next couple of months, Jim quit wearing leather shoes, and he pressured his girlfriend to go veg. He even began to question the morality of keeping pets, including his beloved white cockatiel. One afternoon Jim looked at the bird flitting around her cage in his living room, and a little voice in his head whispered, "This is wrong." Gently, he carried the bird into his backyard. He said good-bye and released the cockatiel into the gray skies of Raleigh, North Carolina. It was a great feeling, he told me. "Amazing!" But then he sheepishly added, "I knew she wouldn't survive, that she probably starved. I guess I was doing it for myself more than for her."

Our relationships with animals can also be emotionally complicated. Twenty years ago, Carolyn fell head over heels for an 1,100-pound manatee. She had applied for a job—any job—at a small natural history museum in central Florida. The museum had an opening; they were looking for a caregiver for a thirty-year-old sea cow named Snooty. Carolyn had no experience working with marine mammals, but they offered her the position anyway. She did not know that her life was about to change.

On the phylogenetic scale, Snooty falls somewhere between the Creature from the Black Lagoon and Yoda. When Carolyn introduced me to him, Snooty hooked his flippers over the edge of his pool, hoisted his head two feet out of the water, and looked me straight in the eye, checking me out. While his brain was smaller than a softball, he seemed oddly wise. I found the experience unnerving. Not Carolyn. She was in love.

For over two decades, Carolyn's life revolved around Snooty. She spent nearly every day with him, even coming around to visit on her days off. Food was a major part of their relationship. Manatees are vegetarians, and Carolyn fed him by hand—120 pounds of leafy green vegetables, mostly lettuce, every day.

But life with an aging sea cow has its downside. Snooty adored Carolyn as much as she doted on him. When she and her husband would sneak off for a week or two of vacation, Snooty would get in a funk and quit eating. All too often, Carolyn would get a call saying that Snooty was off his feed again, and she would rush back to gently ply him with a couple of bushels of iceberg lettuce.

At some point, Carolyn gave up going on vacations. That's when her husband accused her of having her priorities screwed up, of loving a half-ton blob of blubber and muscle more than she loved him.

IS IT WRONG TO FEED KITTENS TO BOA CONSTRICTORS?

As a research psychologist, I have been studying human-animal relationships for twenty years, and I have found that the quirky thinking when it comes to animals that we see in Judith, Jim, and Carolyn is not the exception but the rule. I began to think seriously about the inconsistencies in our relationships with other species one sunny September morning when I got a phone call from my friend Sandy. At the time, I was an animal behaviorist and Sandy was an animal rights activist who taught at my university.

"Hal, I heard that you were picking up kittens from the Jackson County animal shelter and feeding them to a snake. Is it true?"

I was completely taken aback.

"*Arrgh*. What are you talking about? We do have a pet snake, but he is just a baby. He could not possibly swallow a kitten. And I like cats. Even if he were bigger, I would NEVER let him eat a cat."

Sandy apologized profusely. She said she figured the charge was not true, but that she just had to check. I told her I understood, but would appreciate it if she would assure her animal protection pals that I was not dipping into our community's reservoir of unwanted cats to feed my son's snake.

But then I started thinking about the moral implications of keeping a predator for a pet. We had acquired the baby boa by accident. I had spent the summer as a visiting scientist at the University of Tennessee, studying the development of defensive behaviors in reptiles. I was in the lab testing animals one day when the phone rang. It was a stressed-out man who had awakened to find that during the night, his seven-foot red-tail boa constrictor had given birth to forty-two wriggling newborns. He and his wife were understandably shaken; the new mom had never shown any amorous interest in the male with whom she had shared a cage in the couple's living room for the previous eight years.

The man had heard that I was a snake behaviorist and was looking for tips on how to keep the new babies healthy and where he could find good homes for them. I recommended that he contact a reptile expert I knew at the university's veterinary college for information on raising baby snakes, and agreed to adopt one of the babies myself. That evening, my eleven-year-old son, Adam, and I drove to the couple's house, where they gave him the pick of a very large litter. Adam selected the cutest one and named him Sam.

Sam was a low-maintenance pet. He did not scratch the furniture, keep the neighbors awake, or require daily exercise. He was gentle—except for the time he tried to swallow Adam's thumb. It was Adam's fault. He made the mistake of lifting Sam out of his cage immediately after handling a friend's pet hamster. Sam's brain was about as big as an aspirin tablet, and he could not tell the difference between a rodent and a human hand. He just smelled meat.

The accusation that the Herzog family was feeding kittens to snakes came a few weeks later when we were back home in the mountains of western North Carolina. I had no idea how the rumor got started, but the charge, of course, was ridiculous. While boa constrictors are equal-opportunity eaters when it comes to small mammals, Sam was only eighteen inches long and could barely swallow a mouse.

Over the next couple of days, however, several questions kept nagging me. My accuser had inadvertently forced me to confront questions I had never really considered about the moral burdens of bringing animals into our lives. Snakes don't eat carrots and asparagus. Given Sam's need for meat, was it ethical for my son to keep a boa constrictor for a pet? Is having a pet that gets its daily ration of meat from a can of cat food morally preferable to living with a snake? And are there circumstances in which feeding kittens to boa constrictors might actually be morally acceptable?

The person who started the rumor about me lived with several cats that she allowed to roam the woods around her house. Like many cat lovers, she conveniently ignored the fact that from lions to tabbies, all members of the family *Felidae* eat flesh for a living. Each day the cats of America chow down on a wide array of meat. The pet-food shelves of my local supermarket are piled high with six-ounce tins of cow, sheep, chicken, horse, turkey, and fish. Even dried cat foods are advertised as containing "fresh meat." With about 94 million cats in America, the numbers add up. If each cat consumes just two ounces of meat daily, en masse, they consume nearly 12 million pounds of flesh—the equivalent of 3 million chickens—every single day.

In addition, unlike snakes, cats are recreational killers. It is estimated that a billion small animals a year fall victim to the hunting instincts of our pet cats. Oddly, many cat owners don't seem to care about the devastation their feline friends cause to wildlife. A group of Kansas cat owners were informed of the results of a study on the devastating effects of cats on local songbird populations, and then asked if they would keep their cats indoors. Three-fourths of the respondents said no. In a cruel irony, many cat owners also enjoy feeding birds in their backyards, inadvertently luring legions of hapless towhees and cardinals to their deaths at the claws of the family pet. It is likely that at least ten times as many furry and feathered creatures are killed each year as a result of our love of cats than are used in biomedical experiments.

So, pet cats cause havoc. What about pet snakes? Well, first, there are a lot fewer of them. In addition, each snake consumes only a fraction of the flesh that a cat does. According to Harry Greene, a Cornell University herpetologist who studies the feeding ecology of tropical snakes, an adult boa living in a Costa Rican rain forest consumes maybe half a dozen rats a year. This means that a medium-size pet boa constrictor needs less than five pounds of meat a year to stay in good condition. A pet cat requires far more flesh. At two ounces a day, the average cat would consume about fifty pounds of meat in the course of a year. Objectively, the moral burden of enjoying the company of a cat is ten times higher than that of living with a pet snake.

In addition, about 2 million unwanted cats, many of them kittens, are euthanized in animal "shelters" in the United States each year. Presently, their bodies are cremated. Wouldn't it make more sense to make these carcasses available to snake fanciers? After all, these cats are going to die anyway and fewer mice and rats would be sacrificed to satisfy the dietary needs of the pythons and king snakes living in American homes. Seems like a win-win, right?

Yikes . . . I had inadvertently painted myself into a logical corner in which feeding the bodies of kittens to boa constrictors was not only permissible but morally preferable to feeding them rodents. But while the logical part of my brain may have concluded that there was not much difference between raising snakes on a diet of rats or a diet of kittens, the emotional part of me was not buying the argument at all. I found the idea of feeding the bodies of cats to snakes revolting, and had no intention of hitting up the animal shelter for kitten carcasses.

THE PARADOXES OF PET-KEEPING

The boa constrictor incident got me thinking about other instances of morally problematic interactions between people and animals that I had encountered. For instance, my graduate school friend Ron Neibor studied how the brain reorganizes itself after injury. Cats, unfortunately, were the best model for the neural mechanisms he was studying. He employed a standard neuroscience technique: He surgically destroyed specific parts of the animals' brains to observe how their abilities recovered over the succeeding weeks and months. The problem was that Ron liked his cats. His study lasted a year, during which time he became attached to the two dozen animals in his lab. On weekends, he would drive to the lab, release his cats from their cages and play with them on the floor of the animal colony. They had become pets.

His experimental protocol required that he confirm the location of the neurological lesions in the animals in the experimental group by examining their brain tissue. Part of this procedure, technically referred to as perfusion, is grizzly. Each animal is injected with a lethal dose of anesthetic. Then, formalin is pumped through its veins to harden the brain, and the animal's head is severed from the body. Pliers are used to chip away the skull so the brain can be extracted intact and sliced into thin sections for microscopic analysis.

It took Ron several weeks to perfuse all the cats. His personality changed. A naturally cheerful and warm-hearted person, he became tense, withdrawn, shaky. Several graduate students in his lab became concerned and offered to perfuse his cats for him. Ron refused, unwilling to dodge the moral consequences of his research. He did not talk much during the weeks he was "sacrificing" his cats. Killing them took a toll on Ron. Sometimes his eyes were red, and he would look down as we passed in the halls.

These sorts of moral complexities also extend to man's best friend, the dog. My neighbor Sammy Hensley, a farmer who lived just down the road from us on Sugar Creek in Barnardsville, North Carolina, is an example. His two passions were dogs and raccoon hunting. Coon hunting wasn't really a sport for Sammy; it was a way of life. He didn't eat the raccoons he killed. He skinned them out and nailed their pelts and paws to the side of his barn so his neighbors could track his success during the hunting season. (It was while helping him skin a coon that I learned that raccoons—and most mammals—have a bone in their penis; humans are one of the exceptions.) I once accused him of nailing the skins up just to irritate my wife, Mary Jean, who once had a pet raccoon and is nuts about them. But it really wasn't about that. It was just the North Carolina mountain way.

There were two kinds of dogs in Sammy's life—pet dogs and coon hounds—and they led very different lives. He kept four or five hounds at a time, a couple of experienced hunters and a pup or two in training. I loved the names of the breeds: treeing walkers, Plott hounds, blueticks, redbones. Lanky animals with deep voices, languid eyes, oily coats, and the pungent smell hounds have, they usually looked lethargic. That's because they lived most of their lives lying in the dirt, tethered to dog houses by eight-foot chains. But they came alive during hunting season, when they got to tear through the rhododendron thickets in the middle of the night, baying, nose to the ground. You could hear them baying all through the cove.

Sammy loved his hounds. He could tell their voices apart; he knew by the tenor of their yips and yells when they had treed the coon (good) or when they were on a possum's trail (not good). He worried when they got lost and didn't come home in the morning. But they were working dogs, not pets. If a dog couldn't do its job, he would sell it or swap it for a new one.

But Sammy and his wife, Betty Sue, also had pet dogs. While the hounds never saw the inside of the house, the pet dogs—smallish animals like Boston terriers—had the run of the place. Unlike the hounds, these dogs were part of the family. They were petted and played with and allowed to beg for food at the dinner table. One afternoon, when Sammy was mowing hay on a steep section of hillside pasture, his tractor flipped over, killing him. After Sammy died, Betty Sue didn't keep the hounds long, but their little terrier helped her get through the tough times more than anything else. In the Hensley home, the hounds and the pet dogs might as well have been different species.

Most of the dogs living in American homes are simply companions, but our attitudes toward them can be as convoluted as Sammy's relationships with the two categories of dogs in his life. Over half of dog owners think of their pets as family members. A report by the American Animal Hospital Association found that 40% of the women they surveyed said they got more affection from their dogs than from their husbands or children. Yet there is a dark side to our interactions with dogs. One in ten American adults is afraid of dogs, and dogs are second only to late-night noise as a source of conflict between neighbors. (My friend Ross had to sell his house and move because his neighbor's barking dogs turned his life into a nightmare.) In a typical year, 4.5 million Americans are bitten by dogs, and two dozen people, mostly children, are killed by them.

From a dog's eye view, the human-pet relationship isn't always rosy either. Between 2 million and 3 million unwanted dogs are euthanized in animal shelters each year. Then there are the horrendous genetic problems we have inflicted upon dogs in our attempts to breed the perfect pet. Take, for example, the English bulldog, a breed that dog behavior expert James Serpell refers to as a canine train wreck. Bulldogs have such monstrous heads that 90% of bulldog puppies have to be delivered by cesarean section. Their distorted snouts and deformed nasal passages make breathing a chore, even during sleep, and they suffer from joint diseases, chronic dental problems, deafness, and a host of dermatological conditions caused by their wrinkly skin. To add insult to injury, English bulldogs also easily overheat and have a tendency to slobber, snore, fart, and suddenly drop dead from cardiac arrest.

Things are worse for dogs in Korea, where a puppy can be a pet or an item on the menu. Meat dogs, which are typically short-haired, largish animals that look disconcertingly like Old Yeller, are raised in horrific conditions before they are slaughtered, usually by electrocution.

We usually ignore these contradictions but as a psychologist, they began to fascinate me.

FROM THE BEHAVIOR OF ANIMALS TO
THE BEHAVIOR OF ANIMAL PEOPLE

In the weeks after I was accused of feeding kittens to boas, I found myself thinking more about the paradoxes associated with our relationships with animals and less about my animal behavior studies. By conventional standards, my research program was a success. I published articles in good journals, received my share of grant funds, and presented my research at scholarly meetings. But it dawned on me that there were plenty of smart young scientists investigating topics like vocalizations in cotton rats, tool use in crows, and the offbeat reproductive habits of spotted hyenas (female hyenas give birth through their penises). On the other hand, there were only a handful of researchers trying to understand the often wacky ways that people relate to other species. Here was an emerging field, one that I could enter on the ground floor and possibly make a contribution to. Within a year, I had closed up my animal lab to concentrate full time on the psychology of human-animal interactions.

Since shifting from studying animal behavior to studying animal people, my research has largely focused on individuals who love animals but who confront moral quandaries in their relationships with them—the veterinary student who tries not to cry when she euthanizes a puppy, the animal rights activist who can't find someone to date because "just going out to eat becomes an ordeal," the burly circus animal trainer whose life is completely focused on the giant bears he hauls around the country in the dreary confines of an eighteen-wheeler, the grizzled cockfighter who beams when I offer to take a picture of his beloved battle-scarred seven-time winner.

I have attended animal rights protests, serpent-handling church services, and clandestine rooster fights. I have interviewed laboratory animal technicians, big-time professional dog-show handlers, and small-time circus animal trainers. I've watched high school kids dissect their first fetal pigs and helped a farm crew slaughter cattle. I analyzed several thousand Internet messages between biomedical researchers and animal rights activists as they tried—and ultimately failed—to find common ground. My students have studied women hunters, dog rescuers, ex-vegetarians, and people who love pet rats. We have surveyed thousands of people about their attitudes toward rodeos, factory farming, and animal research. We have even pored over hundreds of back issues of sleazy supermarket tabloids for insight into our modern cultural myths about animals. (The original title of our article on tabloid animal stories was "Woman Gives Birth to Litter of Nine Rabbits." Unfortunately, the editor of the journal to which we submitted the manuscript did not find the title sufficiently scientific and insisted we change it.)

Like most people, I am conflicted about our ethical obligations to animals. The philosopher Strachan Donnelley calls this murky ethical territory "the troubled middle." Those of us in the troubled middle live in a complex moral universe. I eat meat—but not as much as I used to, and not veal. I oppose testing the toxicity of oven cleaner and eye shadow on animals, but I would sacrifice a lot of mice to find a cure for cancer. And while I find some of the logic of animal liberation philosophers convincing, I also believe that our vastly greater capacity for symbolic language, culture, and ethical judgment puts humans on a different moral plane from that of other animals. We middlers see the world in shades of gray rather than in the clear blacks and whites of committed animal activists and their equally vociferous opponents. Some argue that we are fence-sitters, moral wimps. I believe, however, that the troubled middle makes perfect sense because moral quagmires are inevitable in a species with a huge brain and a big heart. They come with the territory.

I wrote *Some We Love, Some We Hate, Some We Eat* for anyone interested in human-animal relationships. As a researcher, I normally write for specialists whose job it is to wade through jargon-laden prose that can quickly make your eyes glaze over. But I am convinced that scientists have an obligation to communicate with the public, people who do not know the difference between an analysis of variance and a factor analysis but who are eager to read about current research findings and the hot controversies in our field. The trick is to inform readers about the latest results in a way that is interesting, but at the same time respect the complexity of the issues and be honest about what we know and what we don't.

Many of the topics in the book are controversial. Researchers disagree, for example, about whether your dog feels guilty when it poops on the living room rug; whether children who abuse animals become violent adults; and about the role that meat eating played in human evolution. The passions of the public run high over animal issues such as whether the ownership of pit bulls should be outlawed, or whether trying to discover a cure for cancer is worth the deaths of millions of mice each year. Some of these debates have become bitterly divisive, with the partisans viewing the issues with passion approaching religious zeal. (For this reason, as is customary in ethnographic research, I have changed the names of some of the participants.)

For the most part, I have tried to approach these issues as objectively as I can. This means, of course, that well-intended and intelligent people on both sides of some of these controversies will sometimes disagree with me. That's fine. To this end, I have included an extensive list of research citations and recommended readings at the end of the book. If you want to delve further into the effects of pets on human health or the psychology of animal activism, I point you to some of the relevant studies. My goal is not to change your mind about how we should treat animals but to encourage you to think more deeply about the psychology and moral implications of some of our most important relationships: our relationships with the non-human creatures in our lives.

Late one afternoon in 1986, I was standing in a hallway of a posh Boston hotel deep in conversation with Andrew Rowan, the director of the Center for Animals and Public Policy at Tufts University. We were at one of the first international conferences on human-animal relationships, and we were discussing the paradoxes that so often crop up in our attitudes toward the use of animals. How can 60% of Americans believe simultaneously that animals have the right to live and that people have the right to eat them?

Andrew looked up at me and said, "The only consistency in the way humans think about animals is inconsistency."

This book is my attempt to explain this paradox.

1

Anthrozoology

THE NEW SCIENCE OF HUMAN-ANIMAL INTERACTIONS

> Our failure to study our relationships with other animals has occurred for many reasons. . . . Much of it can be boiled down to two rather unattractive human qualities: arrogance and ignorance.   —CLIFTON FLYNN

The thirty-minute drive from the Kansas City airport to the conference hotel was much more interesting than the three-hour flight from North Carolina. I had flown in for the annual meeting of the International Society of Anthrozoology. I found myself sharing a ride with a woman named Layla Esposito, a social psychologist who tells me she recently completed her PhD dissertation on bullying among middle school children. Puzzled, I ask her why she was attending a meeting on the relationships between people and animals. She tells me that she is a program director at the National Institute of Child Health and Human Development. She is at the conference to let researchers know about a new federal grant program that will fund research on the effects that animals have on human health and well-being. The money is coming from the National Institutes of Health

(NIH) and Mars, the corporate giant that makes Snickers for me and Tempting Tuna Treats for my cat, Tilly. NIH is particularly interested in the impact of pets on children: Is pet therapy an effective treatment for autism? What role does oxytocin (the so-called love hormone) play in our attachment to pets? Are children raised with pets less susceptible to asthma?

"How much money are you giving out?" I ask. Two and a half million dollars a year, she says. "Fantastic! This is just what the field needs," I say. I am thinking that Layla is going to have a very full dance card for the next couple of days.

WHY OUR RELATIONSHIPS WITH ANIMALS MATTER

While $2.5 million is paltry compared to the $6 billion that NIH doles out every year for cancer research, the funds will be a shot in the arm for anthrozoology, a field you have probably never heard of. Anthrozoology is a big tent. It includes the study of nearly all aspects of our interactions with other species. For example, the Kansas City conference included talks on how caring for chronically ill pets affects the quality of lives of their owners; the effect of pet ownership on surviving a heart attack; how children decide whether a strange dog is friendly or dangerous; sex differences in cat behavior (neutered males are more affectionate to humans than are spayed females); and the existence of morality in non-human species.

While animals are important in so many aspects of human life, the study of our interactions with other species has, until recently, been neglected by scientists. Take my field, psychology. For a hundred years, psychologists have concentrated on uncovering behavioral processes such as motivation, perception, and memory, and have neglected important facets of daily life such as food, religion, and how we spend our leisure time. Our relationships with animals, especially our pets, also fall into the category of things that everyday people care about but psychologists usually don't.

One reason behavioral scientists have shied away from studying human-animal interactions is that for many of them the topic seems trivial. This attitude is wrong-headed. Understanding the psychology underlying our attitudes and behaviors toward other species is important for several reasons. About two out of three Americans live with animals, and many people have deep personal relationships with their pets. In addition, our beliefs about how we should treat other species are changing, and a lot of us are torn over whether animals should be used as subjects in biomedical research, or killed because they taste good. The debate over the moral status of animals has become such a divisive social issue that FBI officials have called radical animal rights activism America's greatest domestic terrorism threat. Finally, people are fascinated by anthrozoological research. When I tell someone that I study human-animal interactions, almost inevitably they begin to tell me stories about their wacky dogs or their objections to meat or how their Aunt Sally loves to hunt bears with her Plott hounds.

THINKING LIKE AN ANTHROZOOLOGIST

Anthrozoology transcends normal academic boundaries. Among our numbers are psychologists, veterinarians, animal behaviorists, historians, sociologists, and anthropologists. As in every science, anthrozoologists don't always see eye to eye. We differ in our attitudes toward some of the thorny moral issues that arise in human-animal relationships. We don't even agree on the name of our discipline. (Some prefer to call it human-animal studies.) But, despite these differences, researchers who study our relationships with animals have a lot in common. We all believe that our interactions with other species are an important component of human life and hope that our research might make the lives of animals better.

As academic disciplines go, anthrozoology is a small pond, but in the last two decades we have come a long way. Several journals are devoted to publishing research on human-animal interactions, and the International Society for Anthrozoology holds annual meetings where researchers report their latest findings and argue about whether walking your dog will cause you to lose weight and how long cats have been domesticated. In the United States, courses in human-animal interactions are taught in over 150 colleges and universities, and institutions such as the University of Pennsylvania, Purdue, and the University of Missouri have established anthrozoological research centers.

To get a sense of anthrozoological research, here are a few examples of hot issues in the new science of human-animal interactions. Take, for example, the effectiveness of dolphins as healers, how we select our pets, and the connection between childhood cruelty to animals and adult violence.

DO DOLPHINS MAKE GOOD THERAPISTS?

One of the most important topics in anthrozoology is whether interacting with animals can alleviate human suffering. Animal-assisted therapy (called AAT by anthrozoologists) has been around for decades. The term "pet therapy" was coined in 1964 by Boris Levinson, a child psychiatrist who found that some children who were difficult to work with would open up when they played with his dog, Jingles. The residents in my ninety-two-year-old mother's assisted-living facility perk up when the therapy dogs visit a couple of times a week. I find that spilling my guts to our cat, Tilly, helps me work out my little problems. (Tilly takes a tough love approach to counseling. When I start to whine, she just sniffs and walks away. I would probably do better with a low-energy golden retriever with watery eyes—a canine version of Dr. Melfi, Tony Soprano's shrink.)

But does riding a horse, playing with a dog, or stroking a cat really cure depression or enhance the communication skills of children with autism? Janell Miner and Brad Lundahl of the University of Utah analyzed the results of forty-nine published studies on the effectiveness of AAT in children, adolescents, adults, and elderly people in settings ranging from doctors' offices to long-term residential care facilities. They found that dogs were the most common animal therapists and that AAT was used most often for individuals with mental health problems rather than physical ailments. In most (but not all) of the studies, the subjects did measurably benefit from interacting with their nonhuman therapists. And, on average, the degree of their improvement was about the same as depressed people get from taking drugs like Prozac.

Dolphin therapy, however, is more controversial than AAT involving dogs or horses. Dolphins used for therapy are, after all, wild animals held in captivity against their will. In addition, many of the claims made about the curative powers of dolphins are over the top: Interacting with dolphins, it is alleged, can alleviate Down syndrome, AIDS, chronic back pain, epilepsy, cerebral palsy, autism, learning disorders, and deafness, and can even shrink tumors. Among the presumed healing mechanisms are bioenergy force fields, the high frequency clicks and grunts that dolphins use to communicate with each other, and even the ability to directly alter human brain waves.

Dolphin therapy sounds great. Go swimming, get well. But before you sign up for a couple of weeks in a dolphin tank, you should check out the science behind these claims. Most of them are based on anecdotes, self-reports, or poorly designed experiments conducted by individuals who have a vested interest in the results. Dolphin therapy is particularly attractive to desperate parents who will pay whatever it takes to help their kids with disorders such as autism and Down syndrome. They flock in droves to the more than one hundred therapeutic swim-with-dolphins programs in places like the Florida Keys, Bali, Great Britain, Russia, the Bahamas, Australia, Israel, and Dubai, all of them hoping that, through some unknown force, these creatures with perpetual Mona Lisa smiles will work their magic. Dolphin therapy is expensive. Two weeks at the Curacao Dolphin Therapy and Research Center in the Netherlands Antilles costs roughly 700 bucks for each hour in the water. Is the money well spent? Will their hopes be fulfilled?

Nature does not give up its secrets easily. Scientists have to work hard to get beneath the veil. Just like everyone else, researchers can be duped, particularly when they have a horse in the race. That's why graduate students take courses in research methods and statistics: to learn the tricks of the trade that will help keep them honest. We throw around phrases like "internal and external validity," "placebo control," "random assignment," "single and double blind experiments," and "correlation is not causality." I won't bore you with the details except to say that these conceptual tools help reduce the chances that we will unconsciously tilt the playing field our way.

Good scientists try to be on the lookout for alternative explanations, even if they crush our pet ideas. In 1924, the managers of the Hawthorne Works, a factory outside Chicago, hired a group of psychologists to determine what types of changes in the work environment would make the biggest differences in worker productivity. The psychologists systematically instituted a series of small modifications. First, they increased the lighting on the factory floor, then they made a small change in the pay system. They monkeyed with the work schedule and the length of rest periods. The researchers found that nearly every change they made was followed by a temporary uptick in performance, even when it involved simply undoing a previous change. They concluded that the increases in worker productivity were not due to better lighting or better pay or longer breaks per se. They were just temporary improvements caused by a change in routine.

Could something like the Hawthorne Effect—simply having a new experience—explain the improvements seen in patients undergoing dolphin therapy? Think about it. In addition to hanging out with some of the most appealing creatures on Earth, you travel to beautiful places, spend time floating in tropical seas, and live for a while in a supportive environment where your expectations for success are high.

How can we separate the real effects of interacting with dolphins from all the other neat things that can happen during two weeks at dolphin camp? Fortunately, there are methods to help tease out the actual effects of treatments from those caused by unconscious biases that can creep into our experiments.

In order to take a cold, hard look at whether the benefits of interacting with dolphins are due to more than just temporary feel-good, we need to use a *Consumer Reports*–type approach. What, for example, does the research really show about the effect of ultrahigh-frequency dolphin sounds on handicapped children? A group of German researchers carefully observed sessions in which dolphins interacted with groups of mentally and physically handicapped kids in a dolphin therapy program in the Florida Keys. They found that most of the dolphins ignored the children, and there was not much ultrasonic dolphin talk going on. In fact, the children were exposed to an average of only ten seconds of dolphin ultrasounds during each session, not nearly enough to be beneficial. The researchers concluded that the kids would have been better off playing with dogs.

But what about the dolphins' purported ability to heal through good vibes, a healing smile, and mysterious electric fields? Careful analyses of these claims have been conducted by several researchers. Among them are Lori Marino and Scott Lilienfeld at Emory University. Lori is an animal person. She spends her Saturdays trying to find homes for rescued cats. But her real love is dolphins. She was originally attracted to the unusual anatomy of their brains when she was a graduate student in neuroscience. She has now been studying dolphins for nearly twenty years and was the first scientist to show that they have the ability to recognize themselves in mirrors (a trait shared with humans, apes, elephants, and magpies). Scott is a clinical psychologist who has made a career out of taking on some of psychology's most sacred cows, such as whether those Rorschach inkblots reveal much about your personality (they don't).

Given Lori's expertise with dolphins and Scott's ability to cut through psychobabble, they were the perfect team to assess whether dolphin therapy has a demonstrable effect on troubled bodies and minds. Lori and Scott carefully evaluated the methods of published studies claiming that dolphin therapy is effective for disorders such as depression, dermatitis, mental retardation, autism, and anxiety. They found that every one of them was methodologically flawed: small sample sizes, lack of objective measures of improvement, inadequate control groups, inability to separate the effects of the dolphins from an increased feeling of well-being that comes from doing new things in pleasant environments, and researcher conflicts of interests.

Lori and Scott contend that there is no valid scientific evidence that dolphin therapy is an effective treatment for any of the disorders that its advocates claim. They think it is all pseudoscience. Not content with blowing off dolphin therapy as scientific mumbo jumbo, Lori and Scott want to put the industry out of business. They call it a dangerous fad. I can see the fad part, but why is it dangerous? If you can afford it, why not let kids with too little joy in their lives frolic with Flipper for a couple of weeks? Seems harmless.

Lori doesn't agree. She points out that this "therapy" poses risks for both humans and animals. Dolphins can be aggressive, even to the children they are supposed to be healing. A recent study found that half of over 400 people who worked professionally with marine mammals had suffered traumatic injuries, and participants in dolphin therapy programs have been slapped, bitten, and rammed (the latter resulting in a broken rib and a punctured lung). You can even contract skin diseases from these animal therapists.

Dolphin therapy also raises pesky ethical issues. Clinical psychologists choose to become therapists. Dolphins do not. While most animals used in dolphin therapy programs in the United States are born in captivity, in other countries they are usually captured in the wild, often in massive roundups. Lori says that seven dolphins die for each one that makes it to a cetacean Guantanamo, where it will spend the rest of its life swimming circles in a concrete pool.

Do we have the right to capture intelligent animals with complex social lives and sophisticated communication systems and turn them into therapists for autistic children? I suppose the practice might be justified if these animals really did possess special curative powers. But I would need rock-solid evidence that dolphins can transform the isolated autistic child, or that a couple of hours of dolphin play could add fifteen points to the IQ of a girl with Down syndrome, or that dolphin electric fields could jolt the middle-age depressive out of his debilitating funk. But that evidence does not exist.

Dolphin therapy is an unregulated industry that is not certified or approved by any recognized psychological or medical professional organization. In 2007, the British Organizations the Whale and Dolphin Conservation Society and Research Autism called for a ban on all dolphin therapy programs. Even a pioneer in the dolphin therapy movement has joined the cause. Betsy Smith was an anthropologist at Florida International University in the 1970s when she began bringing dolphins and mentally handicapped children together. At first, the results looked good, and she quickly became a proponent of dolphin therapy. Not any more. In a letter released by the Aruba Marine Mammal Foundation, Dr. Smith wrote that "the primary motive of all captive programs is money." Ouch.

According to my friends who have done it, swimming with dolphins is fun. But marine mammals are not magic bullets. A week of dolphin therapy won't straighten the spine, heal the troubled mind, or prevent epileptic seizures. Save your money; save a dolphin.

DO PEOPLE LOOK LIKE THEIR DOGS?

When people find out I study human-animal relationships, they often tell me, "Oh, you should talk to my friend ____. She is crazy about her ____." When my sister told me I should talk to Paulette Jacobson, I took her up on it. Paulette lives with a Shih Tzu named Miss Bette Davis (Missy for short) on Bainbridge Island near Seattle. Missy is a rescue dog who was severely neglected by her previous owner. Now she lives a life of luxury that includes home-cooked meals, boat rides on Puget Sound, and a fancy wardrobe. Paulette gets a kick out of dressing Missy up. Missy has a raincoat and sweaters, sunglasses and goggles. Sometimes Paulette and Missy dress alike and ride around Bainbridge on their motor scooter. They make a cute couple. People wave and stop to take their picture. A pet boutique is opening on the island, and Paulette can't wait to see the new lines of doggy fashions they will offer. She adores Missy. Paulette told me, "She is everything I want in a dog." But Missy is more than a companion for Paulette. "Missy is my alter-ego. I think of her as a fashion accessory."

Nicole Richie took the idea of her pet being an extension of herself literally when she had hair extensions that matched her own hair color attached to the coat of her dog, Honey Child. That people resemble their dogs is an enduring piece of folk psychology—that's psychology-speak for conventional wisdom. You know the stereotypes—burly bikers with jailhouse tattoos go for pit bulls; leggy fashion models stroll down Park Avenue with pairs of lanky Afghans. But do people really look like their dogs?

University of British Columbia psychologist and dog expert Stanley Coren thought the idea was not that far-fetched. After all, social psychologists have found that people are attracted to romantic partners who are about as attractive as they are. Why shouldn't picking an animal you want to live with follow the same principle? Coren reasoned that if people were attracted to animals that looked like themselves, women with short haircuts that exposed their ears would prefer breeds with sharp prick ears—huskies and basenjis, for example—while women with long hair would prefer floppy-eared breeds like beagles and springer spaniels.

To test his hypothesis, Coren asked women with different hairstyles to rate pictures of four dog breeds that differed in the shape of their ears. Each of the women rated how they liked each dog's looks, how friendly the dog seemed, how loyal it would be, and how smart it appeared. Just as he predicted, Coren found that women with long hair liked the springer spaniels and beagles better and women with short hair preferred basenjis and huskies. In addition, short-haired women rated the prick-eared dogs friendlier, more loyal, and smarter. Coren argued that people like a certain look. They like it on themselves and on the dogs they are attracted to.

Interesting. However, Coren did not actually prove that people tend to look like *their* dogs. That task was taken on recently by psychologists Michael Roy and Nicholas Christenfeld. While reading from storybooks to his kids one night, Christenfeld noticed that dogs in the books often looked like their owners. He wondered if this were the case in real life. And, if so, why?

The researchers came up with two possible reasons why people might look like their dogs—convergence and selection. The convergence theory is that owner and pet actually grow to look more alike over the years. On the surface, the idea seems nutty. However, there is evidence that couples who have been married for a long time do in fact converge in the way their faces look. Plus, obese people tend to have overweight dogs. If the convergence idea is true, the researchers figured that there should be a relationship between how long people live with their dogs and how much they look like them. The selection theory, in contrast, holds that we unconsciously seek animals that look like us when picking a pet. Roy and Christenfeld predicted that if this idea is correct, there should be more dog-owner similarity in purebreds than in mutts. This is because it is harder to know what a mixed-breed puppy will look like as an adult.

To test these ideas, Roy and Christenfeld hung around dog parks and took pictures of owners and their pets. The researchers then made sets from these photographs consisting of a picture of the owner, his or her dog, and a photo of a different dog. They asked college students to try to match the owner with the right dog. If only random chance were operating, the students should make the right match about 50% of the time. But, if the dogs tended to resemble their owners, the judges should do better than that. The researchers thought that the selection theory was a better explanation of owner-dog appearance matching than convergence. Thus they predicted that matching would occur only in purebreds and that there would not be any relationship between how long people lived with their dogs and how similar dog and owner looked.

The researchers were right on all counts. The students correctly matched owners and their purebred dogs two-thirds of the time. This was a significantly better hit rate than would be expected if they were just randomly guessing. And, just as predicted by the selection theory, the students were not successful at matching owners of mixed breeds with their pets. Finally, as predicted by their selection theory, there was no evidence that people grew to look more like their pets the longer they lived with them.

Like most scientists, I have a skeptical streak. I was not at all convinced when I first read Roy and Christenfeld's article. But I have become a believer. Research groups in Venezuela, Japan, and England subsequently found that people can match pictures of owners and their dogs at better than chance levels. While not everyone resembles their pets, the scientific support for the idea that a lot of people do tend to look like their dogs is surprisingly strong. Go figure.

DO DOG PEOPLE AND CAT PEOPLE HAVE
DIFFERENT PERSONALITIES?

My friends Phyllis and Bill have a mixed marriage. She is a cat person; he is not. Phyllis has had cats since she was in college, usually two or three at a time. I once spent a month housesitting for her and agreed to give one of her cats, the foul-tempered Chris, two pills every day, one for epilepsy, the other for depression. It was a daily struggle that I won only part of the time. In recent years, Phyllis has forked out thousands of dollars to veterinarians for patching up Chipper, her gray tabby, who has a penchant for going into full battle with stray tomcats and raccoons.

What does Phyllis like about cats? She claims it's their nicely balanced need for both affection and independence, a mix that she also likes in her husband. She thinks dogs are suck-ups.

Bill, on the other hand, does not particularly like animals. He never has. His parents didn't have pets when he was growing up, and Bill never felt the slightest desire to live with one himself. But then he married Phyllis and he was suddenly living with cats. Over time, his attitudes toward the cats in his home have shifted slightly, from indifference to tolerance. He admits that he enjoys letting one of them lie on his belly at night when he watches the news. But he doesn't feed the cats and he never asks how they are doing when he talks to Phyllis on the phone when he is away. Bill says that if he were living by himself, he would not have a pet at all.

Phyllis is a psychotherapist, a good one. Given her clinical expertise, I asked if she saw a difference between cat people and dog people. I was surprised when she said no, that it was not personality but serendipity that determines the types of pets people fall for. A cute kitten simply shows up in your backyard or you happen to grow up in a family that has dogs or you want an animal companion that will get rid of the mice in your basement.

I am almost certain that you think of yourself as either a dog person or a cat person, probably a dog person. That's because, if asked, most people will instantly put themselves into one of these categories. And according to a recent Gallup poll, 70% of Americans say they are dog people. This aspect of pet demography is paradoxical as there are more cats than dogs in American homes. (Mary Jean and I, by the way, are both dog people, even though we live with a cat.)

But is it true that dog people and cat people have different personalities, or is this yet another piece of common sense that proves to be wrong?

This question was taken on by Sam Gosling, a psychologist at the University of Texas who studies individual differences in people and in animals. His research on human personality (which is described in his fascinating book, *Snoop: What Your Stuff Says About You*) has shown that while some of our personal preferences reveal aspects of our personality traits, others do not. He can, for example, tell a lot about you from knowing what music is loaded on your iPod, how messy your bedroom is, and whether you hang inspirational posters on your office wall. On the other hand, he has found that the contents of your refrigerator say nothing about what you are really like.

But before we can answer the question of dog people versus cat people, a brief lesson in the psychology of personality is in order. Psychologists have been arguing about the nature of human personality for a hundred years. One issue they fight about is how many personality traits there are. While there are a few holdouts, most psychologists agree that we can get a good description of a person's personality by measuring five basic traits. (Technically this is referred to as the Five Factor Model; psychologists usually just call it the Big Five.)

The Big Five traits are:

- Openness Versus Closed to Experience
- Conscientiousness Versus Impulsiveness
- Extroversion Versus Introversion
- Agreeableness Versus Antagonism
- Neuroticism Versus Emotional Stability

Sam and an anthrozoologist at Cambridge University named Anthony Podberscek wondered if the personalities of pet owners were different from non–pet owners. They scoured the scientific literature, located dozens of studies comparing the two groups, and found a hodgepodge of results. For every study reporting that pet owners were more extroverted or more emotionally stable or less independent than non–pet owners, there was another one that found no difference between the two groups. They concluded that there was no evidence that pet owners were different from non–pet owners in their basic personalities.

Is this also true of the dog person/cat person dichotomy? Sam maintains an online version of the Big Five Personality Test that thousands of people have taken over the last ten years. (You can take it yourself by going to www.outofservice.com/bigfive.) In 2009, he temporarily added an item in which participants were asked if they considered themselves to be a dog person, a cat person, neither, or both. In a little over a week, 2,088 dog people and 527 cat people had taken the personality test.

Here are the results:

- Dog people are more *extroverted*.
- Dog people are more *agreeable*.
- Dog people are more *conscientious*.
- Cat people are more *neurotic*.
- Cat people are more *open to new experiences*.

So, in this case, folk psychology is right—there is a difference between dog people and cat people, and most of the differences are along the lines that you probably would have predicted. But in science, there is often a catch. In this case, the catch is that the differences in their personality scores were relatively small. (The exception was extroversion, which was in the moderate size range). The bottom line is that whether you call yourself a dog person or a cat person tells us something about your personality—not as much as the contents of your iPod, but more than the state of your refrigerator.

DO CHILDREN WHO ABUSE ANIMALS
BECOME VIOLENT ADULTS?

On a recent visit to Manhattan, I spent an afternoon strolling through the Metropolitan Museum of Art looking for paintings depicting human-animal relationships. There were lots of them, but one of the most striking was an oil painting by a sixteenth-century Italian artist named Annibale Carracci aptly titled *Two Children Teasing a Cat*. The painting portrays an innocent-looking young boy and girl and a cat. The boy is holding the cat with his left hand and a large crayfish in his right. He has provoked the crayfish into clamping one of its massive claws onto the cat's ear. That the children are angelically smiling, apparently delighted with their "game," makes the painting especially chilling. What should we make of this wanton cruelty? Is it just a childish prank or an indicator of deep-seated psychopathology that will someday erupt into far worse violence?

The infliction of abject cruelty toward members of other species illustrates how our interactions with animals reflect larger themes in psychology. For example, is animal abuse the result of nature or nurture? Some scientists believe that roots of cruelty lie in our evolutionary history, particularly the fact that our ancestors were likely meat-eating apes that delighted in ripping their prey to pieces. Others, however, argue that human children are naturally kind, and that callousness toward animals is instilled in us by a culture that promotes activities like hunting and eating flesh. Cruelty also offers fodder for those looking for moral inconsistencies in our treatment of animals. What, for instance, is the moral difference between the pleasure that a hunter derives from shooting deer and that a mean child gets from tying a tin can to the tail of a dog?

The anthropologist Margaret Mead wrote that "one of the most dangerous things that can happen to a child is to kill or torture an animal and get away with it." She was echoing a theme that has been knocking around for hundreds of years. John Locke and Immanuel Kant connected animal cruelty and human-directed violence. In fact, Kant thought that the only reason we should be nice to animals is that animal cruelty leads to human-directed brutality. Some anthrozoologists are convinced that animal abuse in children is often the first step on a path that leads to adult criminality. Others, however, are not so sure.

One of the first systematic studies associating animal cruelty and criminality was conducted by Alan Felthous, a psychiatrist, and Stephen Kellert, a leader in the study of human-animal interactions. They interviewed groups of aggressive criminals, nonaggressive criminals, and noncriminals. The highly aggressive criminals were much more likely to repeatedly abuse animals than men in the other groups. And their level of violence was different. They cooked live cats in microwave ovens, drowned dogs, and tortured frogs.

In the wake of this and similar studies, I have taken to asking my friends if they ever abused animals when they were children. It was an eye-opener. For example, my buddy Fred, a builder, confessed that he and his childhood pals blew up frogs with firecrackers. When he was five, Henry's mom bought him a little brown puppy with floppy ears. One day Henry and his friends decided to play catch with the puppy by tossing it back and forth over a picket fence. The dog banged into the pickets over and over. The pup died a couple of days later. Henry told me that just thinking about it now makes him want to cry. When I asked Linda if she had participated in animal cruelty as a child, she got very quiet and suddenly serious. She said yes but that she just could not talk about it. Ian was the least of the offenders. All he did was fry ants with a magnifying glass.

I was surprised that so many people I knew admitted to abusing animals when they were little. Yet none of them turned to the dark side—no felons, wife-beaters, or serial killers among them. Neither did Charles Darwin, who wrote in his autobiography that, as a boy "I beat a puppy, I believe simply from enjoying the power." (However, Darwin then wrote, "This act lay heavily on my conscience as shown by my remembering the exact spot where the crime was committed.")

I plead guilty, too. When I was growing up in Florida, my friends and I used land crabs and toads for target practice with our Daisy Red Ryder BB guns. One morning, BB gun in hand, I saw a songbird sitting on a limb. I figured, why not take a shot at it? I was sure I would miss. And, after all, a BB wouldn't do much damage. I was wrong. A puff of air, and the bird silently dropped to the ground, dead. I was horrified. There was a huge gulf between an ugly land crab and a lively bird perched in a tree. It was the last animal I ever shot.

The idea that there is a strong link between childhood animal cruelty and violence directed toward people is so well established that the term "The Link" is now a registered trademark owned by the American Humane Association. Public presentations by Link advocates often begin with tales of tragedy. First, the serial killers: Albert DeSalvo (the Boston Strangler), Jeffrey Dahmer, Lee Boyd Malvo (the D.C. sniper accomplice)—all of them accused of childhood animal cruelty. Then the school shootings: Columbine, Colorado; Springfield, Oregon; Jonesboro, Arkansas; Pearl, Mississippi; Paducah, Kentucky—again, all committed by boys reputed to have a history of animal abuse.

I have never been overly impressed with this type of anecdotal evidence. Some Link advocates would have you believe that most or even all serial killers and school shooters had a history of childhood animal abuse. Not true. An analysis of 354 cases of serial murders found that nearly 80% of the perpetrators did not have a known history of cruelty to animals. The connection between school shootings and animal cruelty is even more tenuous. In 2004, a joint task force of the U.S. Secret Service and the Department of Education undertook a thorough examination of the psychological characteristics of the perpetrators of thirty-seven school shootings. The researchers found that only five of the shooters had a history of animal abuse. They concluded, "Very few of the attackers were known to have harmed or killed an animal at any time prior to the incident." Clearly, some Link proponents overstate the relationship between childhood animal abuse and adult violence. There is, however, some evidence that the two are somehow related. The problem is determining how close the relationship is and why it exists.

There are several reasons why you might think that animal cruelty in children and later violence would be connected. I call the first one the bad seed hypothesis. Some children are liars, cheats, thieves, and bullies by the time they are in elementary school. Psychiatrists refer to this pattern of behavior as a conduct disorder. In the 1960s, it was thought that three traits were especially characteristic of these kids: fire-setting, bedwetting, and animal cruelty. While this trio is not as closely connected as originally thought, the American Psychiatric Association still includes animal cruelty as a diagnostic criterion for conduct disorders. The bad seed hypothesis holds that animal abuse is not a cause of delinquency, but a sign of a seriously troubled child, many of whom will become psychopathic adults.

A stronger version of Link thinking is called the violence graduation hypothesis. This is the idea that pulling the wings off butterflies or beating a dog is the first in a series of steps that can eventually end in prison. The title of an influential book by Linda Merz-Perez and Kathleen Heide captures this idea—*Animal Cruelty: Pathway to Violence against People*. One implication of this theory is that animal abuse in children can be used as a form of profiling, a way of identifying the potential serial killers and school shooters before their violence escalates.

So, are the data in? Can we conclude that childhood animal cruelty causes later violence? Not necessarily. A group of researchers led by Arnold Arluke, a sociologist at Northeastern University, came up with an innovative way to test the graduation hypothesis. They compared the criminal records of individuals who had been convicted of animal abuse with a group of law-abiding citizens from the same neighborhoods. The researchers reasoned that if the graduation hypothesis were true, the animal abusers would show a propensity for violent crimes rather than run-of-the-mill offenses such as selling drugs or auto theft.

Their results did not support the graduation hypothesis. True, the animal abusers were big-time troublemakers. You would not want to live next door to any of them; they committed many more crimes than the noncriminals they were matched with. But they were equal-opportunity bad guys. They were no more likely to be arrested for violent crimes than they were to be arrested for nonviolent offenses like burglary or peddling drugs.

There are other reasons we should be careful about making causal connections between childhood cruelty and adult violence. In Philosophy 101 (Logic), you learn that "all As are Bs does not imply that all Bs are As." Thus the fact that most heroin addicts first smoked marijuana does not imply that most first-time marijuana smokers will become junkies. Similarly, even if every school shooter and every serial killer had abused animals as a child (a dubious claim), we could not logically conclude that children who pull the wings off moths are apt to become murderers.

More important, the numbers do not support the idea that most early animal abusers become violent when they grow up. Emily Patterson-Kane and Heather Piper analyzed the results of two dozen research reports of childhood cruelty among extremely violent men (serial killers, sexual abusers, school shooters, rapists, and murderers) and males with no history of violence (college students, teenagers, and normal adults). They found that 35% of the violent offenders had been childhood animal abusers—but so had 37% of the males in the "normal" control group. Suzanne Goodney Lea, a sociologist, obtained similar results. She studied the backgrounds of 570 young adults, 15% of whom had a history of animal abuse. She found that children who got in fights, lied habitually, used weapons, or set fires did tend to become violent adults. Animal cruelty, however, did not predict later aggressive behavior.

Arnold Arluke has a talent—listening to people. He puts them at ease, and they tell him things they would not normally reveal. He would have been a good homicide detective. Arluke used this ability to delve into the minds of college students who had tortured animals. They weren't hard to find. He just asked students in his classes. The students he interviewed had poisoned fish with bleach, ripped the legs off flies, burned grasshoppers with lighter fluid, and played Frisbee with live frogs. This statement by a woman he interviewed is typical: "It was like we didn't have anything to do and we were bored, so it's like, 'OK, let's go torture some cats!'"

Arluke's students were not an anomaly. In one recent study of college students, 66% of male students and 40% of female students admitted that they had abused animals. Arluke has come up with a radical suggestion. He believes that for many children, animal cruelty is a normal part of growing up. He calls it dirty play. It's forbidden fruit, like swearing or smoking cigarettes. He thinks animal abuse enables children to play adult power games in secret. It is also helps cement bonds between the co-conspirators, your partners in crime. Granted, the types of childhood cruelty that Arluke uncovered in his presumably normal sociology students were generally not of the microwaving cats and dropping puppies off the roof variety. And, unlike the hard-core criminals interviewed by Felthous and Kellert, most of Arluke's students felt remorse for their youthful indiscretions. But the fact remains that childhood animal abuse is more common than is generally recognized.

The awkward fact is that most wanton animal cruelty is not perpetrated by inherently bad kids but by normal children who will eventually grow up to be good citizens. For me, animal abuse raises a big question, but it is not why deranged psychopaths are cruel. The answer to that question is obvious; they are mentally ill, morally blind, or evil. No, the more interesting and important issue transcends our relationships with animals: Why do fundamentally good people do fundamentally bad things?

For some researchers and animal protection organizations, the connection between animal cruelty and human violence has become a moral crusade pursued with missionary zeal. Some researchers, however, have come to question simplistic Link thinking. They worry that Link advocates and the media are perpetuating an irrational moral panic among the public. Link skeptics don't argue that we should ignore animal abuse. Rather, they believe that we should treat animal abuse as a serious problem in its own right, not because it turns children into adult psychopaths.

Anthrozoologists are divided over the strength of the connection between animal abuse and human violence. This debate is no different than contentious issues in other areas of human behavior. For years, psychologists have argued over whether violent TV causes aggression in children, whether pornography fuels sex crimes, and whether day care is good or bad for children. Like these questions, the controversy over the causes and effects of animal cruelty is not going away. The issue is too important.

————

As you can see from these examples, the scope of anthrozoology is broad. We investigate issues like personality differences between cat and dog people simply because they are interesting. On the other hand, studies of the effectiveness of animal-assisted therapy and the link between animal cruelty and adult aggression have important real-life implications. But there is another reason that the study of our interactions with other people is both fascinating and important. It is that our interactions with animals offer an unusual glimpse into human nature. As I show in the next chapter, the French anthropologist Claude Lévi-Strauss got it right when he wrote, "Animals are good to think with."

2

The Importance of Being Cute

WHY WE THINK WHAT WE THINK ABOUT
CREATURES THAT DON'T THINK LIKE US

It's easier to empathize with the dog than with the flea.   —ERIC GREENE

Judy Barrett of Greensboro, North Carolina, had a problem. She and her husband were crazy about bluebirds. They had spent a lot of money to entice a pair of them to nest in their backyard, even purchasing a snake-proof bluebird nest box and special bird baths. Judy kept a supply of mealworms in her refrigerator because she heard that bluebirds love them. The Barrett's amenities did attract a pair of birds, but not the kind they wanted. When they weren't looking, a common sparrow hijacked the nest box and laid five eggs in the bluebird house.

Not knowing what to do, Judy sent a letter to Randy Cohen, who writes "The Ethicist," a Dear Abby–style advice column on everyday moral problems in the *New York Times Magazine*.

Would it be ethical, Judy asked, for her to destroy the eggs of the lowly sparrow to make room for a pair of lovely bluebirds?

Cohen said no. "In ethics, cuteness doesn't count."

Logically, of course, he is right. But while cuteness may not count in the rarified world of moral philosophy, it matters a lot in how most people think about the treatment of other species. For instance, one of the biggest factors in how much money people say they would donate to help an endangered species is the size of the animal's eyes. This is bad news for the rare giant Chinese salamander. It is the largest and possibly most repulsive-looking amphibian on Earth, a beady-eyed six-foot-long mass of brown slime. You don't see pictures of them gracing the fund-raising brochures of environmental organizations. But contrast the giant salamander with another Chinese animal, the equally rare but infinitely more appealing giant panda, whose eyes are exaggerated by giant dark circles. They are so endearing that the panda is the logo of the World Wildlife Fund.

Of the 65,000 species of mammals, birds, fish, reptiles, and amphibians on the planet, only a handful merit much human concern. Why do we care about the giant panda but not the giant salamander, the eagle but not the vulture, the bluebird but not the sparrow, the jaguar but not the Dayak fruit bat (one of only two species of mammals in which males produce milk)? The ways that we think about animals are often determined by species' characteristics—how attractive the creatures are, their size, the shape of their head, whether they are furry (good) or slimy (bad), and how closely they resemble humans. Too many legs or not enough legs are negatives. So are disgusting habits like eating feces or sucking blood. How an animal's flesh tastes also counts, though not as much as you might think.

The inconsistencies that haunt our relationships with animals also result from the quirks of human cognition. We like to think of ourselves as the rational species. But research in cognitive psychology and behavioral economics shows that our thinking and behavior are often completely illogical. In one study, for example, groups of people were independently asked how much they would give to prevent waterfowl from being killed in polluted oil ponds. On average, the subjects said they would pay $80 to save 2,000 birds, $78 to save 20,000 birds, and $88 to save 200,000 birds. Sometimes animals act more logically than people do; a recent study found that when picking a new home, the decisions of ant colonies were more rational than those of human house-hunters.

What is it about human psychology that makes it so difficult for us to think consistently about animals? The paradoxes that plague our interactions with other species are due to the fact that much of our thinking is a mire of instinct, learning, language, culture, intuition, and our reliance on mental shortcuts.

BIOPHILIA: IS LOVE OF ANIMALS INSTINCTIVE?

In an elegant little book written in 1984, the Harvard biologist E. O. Wilson hypothesized that our species has an instinctive affiliation with the natural world. He called this trait biophilia and argued that it is a built-in part of human nature. Though I was initially skeptical of the idea, evidence is amassing that he is right. The developmental psychologists Judy Deloache and Megan Picard have discovered that even very young infants pay more attention to films of real animals than they do to films of inanimate objects. A team of evolutionary psychologists from the University of California, Santa Barbara, have shown that the human visual system is particularly adept at picking out animals in the environment, a capacity that would have served our ancestors well as they needed to be on the lookout for both predators and prey. They call this idea the animate-monitoring hypothesis, and their experiments show, for example, that people are quicker to spot the movements of an elephant than those of a truck.

Our attraction to some animals does seem to be instinctive. When I give talks about human-animal relationships, I usually include a couple of slides that inevitably evoke a chorus of *oohs* and *ahhhhs* from the audience. The pictures are of kittens and puppies. The audiences' responses to the photographs reflect a component of human nature that makes most behavioral scientists squirm: instinct. The notion that humans are innately drawn to anything that looks like a baby—infants, puppies, ducklings, you name it—is called the "cute response." The idea was first proposed by the Austrian ethologist Konrad Lorenz. Young animals share features with human infants; large foreheads and craniums, big eyes, bulging checks, and soft contours. Lorenz referred to these characteristics as "baby releasers" because they automatically bring out our parental urges.

Bambi is the classic example of how easily we are manipulated by baby releasers. Walt Disney originally urged the animators working on the film to draw the fawn as accurately as possible. He had a pair of fawns shipped in from Maine and made his artists watch an anatomist dissect the rotting carcass of a newborn deer. The problem was that the Bambi sketches that the animators produced, while realistic, were not cute enough to grab the hearts of the movie-going public. The solution was babyfication; Disney told the artists to reduce the length of Bambi's muzzle and make Bambi's head bigger. Then they gave Bambi huge eyes with lots of white in them. Bambi was morphed into a surrogate human baby.

Mickey Mouse is a similar testament to Disney's ability to design characters that elicit our parental urges. Mickey started life in 1928 as a not-so-nice trickster named Steamboat Willie. Over the next fifty years, Disney systematically changed his image. To accomplish this shift to a kinder and gentler Mickey, his features became more baby-like. Mickey's head grew to nearly half the size of his body, and the size of his eyes and brain case nearly doubled. Does our innate tendency to be taken in by a pair of oversized eyes affect our attitudes toward the treatment of other species? Of course. Stephen Jay Gould, the late Harvard biologist who traced Mickey's evolution said it best: "We are, in short, fooled by an evolved response to our own babies and we transfer our reaction to the same set of features in other animals."

The role of cuteness in our attitudes toward animals is illustrated by public outrage over the annual "harvest" of baby harp seals on the ice floes off the Atlantic coast of Canada. The seals are irresistible right after they are born; for the first two weeks of their lives, their fur is pure white and their eyes dark and as deep as pools. In the 1970s and 1980s, gory photographs showing the oozing blood of newborns being clubbed to death were staples of the brochures and placards of anti-hunt protesters. In 1987, the Canadian government caved in to public pressure—sort of. They prohibited killing seal pups under fourteen days old, which happens to be when their fur becomes darker and the animals begin to look less infantile. Then it is open season. The Canadians did not stop the baby seal hunt; they stopped the *cute* baby seal hunt.

Our fetish for animals that look like infants comes at a cost. Humans' love for the cute has produced canine breeds in which full-grown dogs resemble perpetual puppies. The babyish snouts of breeds like Chinese pugs and French bulldogs make for respiratory problems, and their bulging puppy eyes barely fit into their shallow sockets. By breeding dogs for neoteny (the biological term for the retention of juvenile features in adults), we have also created pets that are emotionally immature and prone to canine versions of our own neuroses. This phenomenon has been a boon to Big Pharma, which has developed repackaged versions of Valium and Prozac for our depressed, anxious, and obsessive-compulsive pets.

WHY DO PEOPLE HATE SNAKES?

But if people are *biophilic* toward creatures like puppies and baby seals, they are *biophobic* toward others—snakes, for instance. In a 2001 Gallup poll, Americans were asked about the things that make them sweat. Four of their top ten fears were of animals, with snakes at the top of the list. (The other common animal fears were of spiders, mice, and dogs.) Even the revered medical missionary Albert Schweitzer, whose philosophy emphasized reverence for all life, kept a gun around to shoot snakes.

As a young researcher, I observed first hand the conflict between the fascination for snakes and the fear of them. I was spending the summer at a reptile theme park in Florida, recording the mating calls of alligators. At the beginning of the tourist season, the park would hire college students who had experience handling reptiles to lead tours of the facility. At the end of each tour, the guide would don a pair of snake-proof boots and wade into a pit containing dozens of big rattlesnakes and water moccasins, snakes that can kill you.

The balloon trick was the climax of the show. The tour guide would blow up a balloon and pick out a big diamondback, which he would harass with a snake hook until the animal was coiled and ready to strike. To pull off the trick, you have to hold one end of the balloon in your hand and slowly push the other end toward the hot snake. Then, when the balloon is a foot from the snake's nose, you make a quick thrust, pushing the balloon directly into its face. If you do it right, the snake hits the balloon full-force with its fangs. Bang. The balloon bursts. The startled tourists jump and clap and maybe even give you a tip.

But one of the college boys did not have the guts to make the quick final thrust toward the inch-and-a-half fangs of a rattlesnake. The old-timers on the staff did not think much of the summer college kids, especially this one. In the mornings, before the place officially opened, I would join them around the pit to watch the new kid try to learn the balloon trick. Looking confident and cocky in his starched khaki jungle shirt, he would enter the pit, pin a snake, grab its head, and milk it by hooking the snake's fangs over a glass vial and massaging its venom glands. No problem. But then it was time for the finale, the balloon trick. You could see his hands start to tremble when he started to blow up the balloon, and the tremors would get worse as he picked out his target, an Eastern diamondback.

That's when the old guys would start on him, some of them clucking like chicken, a few whispering encouraging words: "Come on, kid . . . you can do this." Then the college boy would line up the balloon and start to push it slowly toward the snake. But slow doesn't do it for a rattlesnake. You've got to be quick to make them strike. You've got to startle them.

The college kid would cautiously edge the balloon closer and closer to the snake's face until it touched the animal's nose, pushing the snake backward out of its strike coil, making a rattlesnake packing enough venom to kill five men look about as tough as a pussycat. Not a good way to impress bloodthirsty tourists.

The kid, humiliated, would leave the pit, eyes down, the old timers clucking, showing no mercy. On day seven of snake tour-guide training, the kid did not show up for work, and I never saw him again. The incident reminded me of the biblical warning that the spirit might be willing, but the flesh is weak. On those morning sessions at the reptile park, primordial fears of the flesh prevailed.

Objectively, the fear of snakes among Americans does not make sense. There are only about a dozen snakebite deaths a year in the United States, and most victims are testosterone-fueled males with more balls than brains.

A case in point was described in an article in the *Annals of Emergency Medicine*. A forty-one-year-old man showed up at a hospital emergency room with a rattlesnake bite on the end of his tongue. The report speaks for itself: "A friend held the snake close to the patient's face while the patient mimicked the tongue protrusions of the reptile. Seizing the opportunity, the rattlesnake quickly bit him on the dorsal surface of the tongue. While the fangs were still in place, the friend yanked the snake out of the patient's mouth." Ouch. The man's tongue swelled to the size of an orange, making it nearly impossible for him to breathe, and he almost died.

Why are so many Americans afraid of snakes? After all, you are more likely to be killed by a dog than by a snake bite. Is ophidophobia a relic of Bronze Age myths featuring serpents, naked women, and apples? Or are people weirded out by the snake's alien leglessness or its phallic form? Or did snake phobias evolve because they steered our ancestors away from animals that could kill them?

Scientists have been arguing for two hundred years about the relative importance of nature and nurture in development of snake fears. Susan Mineka, a psychologist at Northwestern University, argues that in monkeys, fear of snakes is learned. She found that Rhesus monkeys captured in the wild were terrified of snakes but that monkeys born in captivity showed no fear of them. However, if lab-reared monkeys that have never seen a snake observe how its wild-caught brethren react to them, they immediately become snake-phobic.

Other researchers, however, do not think that primates are blank slates when it comes to snakes. The University of Tennessee ethologist Gordon Burghardt and his colleagues at the Kyoto Primate Institute tested adult captive Japanese monkeys in a situation in which they had to reach in front of a snake cage to get food. Many of the animals were absolutely terrified of the snakes even though they had never seen one before. In her book, *The Fruit, the Tree, and the Serpent: Why We See So Well*, Lynne Isbell of the University of California at Davis makes a convincing case that the primate brain was shaped by evolution to specialize in visually detecting snakes. University of Virginia psychologists Vanessay LoBue and Judy DeLoache (the latter of whom has a snake phobia herself) tested the idea that humans have a built-in snake detector. They asked, young children who had never seen a snake to pick out photographs of serpents embedded in a series of pictures of other natural objects. Sure enough, their subjects were quicker at spotting a picture of a snake amid pictures of other animals than they were to pick out a picture of a flower or a centipede.

So nature plays a role in snake fears. But that can't be the whole story. About half of Americans say they are not afraid of snakes, and 400,000 people in the United States keep them as pets. Further, cultures differ in how they treat snakes. My friend Bill spent five years in Tanzania as a game warden. In the village where he lived, people did not distinguish between poisonous and harmless snakes. When anyone saw a snake, they would shout "Nyoka!" and everyone would come running and help club it to death. But this is not true in New Guinea. According to the biologist Jared Diamond, New Guineans are not afraid of snakes despite the fact that a third of snake species on the island are highly venomous. Unlike Tanzanians, New Guinea tribesmen are adept at telling poisonous from nonpoisonous species, and they eat the harmless ones.

The idea that both genes and environment influence our attitudes toward animals fits nicely with E. O. Wilson's updated view of biophilia. He originally conceived of biophilia as a hard-wired human instinctive urge to affiliate with all things bright and beautiful. A few years later, however, he revised the concept to include the profound effects that learning has on our relationships with nature. "Biophilia," he wrote, "is not a single instinct but a complex of learning rules that can be teased out and analyzed individually." And the job of teasing out the learning rules that govern our relationship with nature will fall within the province of anthrozoology.

WHAT'S IN A NAME? LANGUAGE AND MORAL DISTANCING

How we think about animals is also affected by the names we give them and the words we use to describe them. Animal words permeate human language. Some of them elevate ("busy as a bee," "foxy lady!"), others demean ("you bitch"), and some reflect sexual power (*cock, pussy*). Calling someone "an animal" reflects our ambivalence about our place in nature. In some contexts, it is a compliment; in others, an insult. Psycholinguists argue about whether language reflects our perceptions of reality or helps create them. I am in the latter camp. Take the names we give the animals we eat. The Patagonian toothfish is a prehistoric-looking creature with teeth like needles and bulging yellowish eyes that lives in deep waters off the coast of South America. It did not catch on with sophisticated foodies until an enterprising Los Angeles importer renamed it the considerably more palatable "Chilean sea bass."

The words we use for meat help us avoid thinking about the ethical implications of our diet. It is easier to order a pound of beef from the butcher than a pound of cow. Semantic moral distancing is apparently less necessary as we descend the phylogenetic scale; we don't bother with linguistic cover-ups for chicken, duck, or fish. In other parts of the world, however, people dispense with meat euphemisms altogether. The German words for pork, beef, and veal are, respectively, *Schweinefleisch* (pig flesh), *Rindfleisch* (cow flesh), and *Kalbfleisch* (calf flesh). In Mandarin, beef is *niurou,* which translates into cow (*niu*) meat (*rou*); pork is *zhurou* (pig meat), and mutton is *yangrou* (sheep meat).

Partisans on both sides of the animal rights debate realize the power of words. In describing the Canadian seal hunt, the government agency that oversees the hunt uses neutral words: "harvest," "cull," "management plan." The language of seal hunt opponents is peppered with hot words: "slaughter," "massacre," "atrocity." What the wildlife managers call the "swimming reflex of dead animals," the activists refer to as "being skinned alive."

The animal rights group People for the Ethical Treatment of Animals (PETA) has made millions of Americans aware of the suffering associated with factory farming, hunting, animal research, and zoos and circuses. But they have had almost no success in riling up the public over the suffering caused by our insatiable desire for sushi-grade bluefin tuna or the pain experienced by a sixteen-inch brown trout who mistakes a #14 dry fly for a real insect. My friend Cathy says she never eats anything with a face, but she doesn't count fish. PETA's new strategy to change the way we think about creatures with fins rather than fur is to rename them. The slogan for their new campaign against fishing: "Save the Sea Kittens!"

Joan Dunayer would approve. The author of *Animal Equality: Language and Liberation,* she believes that some words makes it easier for us to exploit other species. She proposes linguistic substitutions such as "aqua-prisons" for aquariums, "inmates" for zoo animals, and "cow abusers" for cowboys. She wants us to refer to our pets as "my dog friend" or "my cat friend." I am happy to call Tilly "my cat friend," but I suspect that my dentist, who has an aquarium in his waiting room, will be reluctant to say that it is time to change the water in his "fishy friends' aqua-prison."

PETS OR RESEARCH SUBJECTS? CATEGORIES COUNT

The language that we use to talk about animals is closely tied to another factor that affects how we think about animals—the categories we put them in. For example, animals in the category "pet" are named; animals in the category "research subject" are usually not. When I recently asked a biologist if any of the mice in his lab had names, he looked at me as if I were crazy. I wasn't surprised. After all, the white mice he pokes, probes, and injects are essentially identical. Why should they merit names?

But sometimes our animal categories become blurred. When I was a graduate student we did give names to some of our animal research subjects—the lifers, who became more pets than objects to be used in experiments. Our favorite lab animal was a spectacular five-foot black rat snake named IM (pronounced *em*). We got him when he was just a baby. IM was unusual in that he had two heads and one penis (most snakes have one head and two penises). One head was named Instinct and the other one Mind. You can see why we gave him a nickname.

But the shift from lab experimental subject to pet can come at a cost. A laboratory animal veterinarian told me about the time she "instantly fell in love" with a beagle puppy that was scheduled to be part of an experiment that would end in the animal's death. She quietly took one of the lab technicians aside and told him to swap animals, and another dog was euthanized in place of the beagle. She realized that the beagle had lived only because a person of authority (her) had taken a shine to it, and several years later she still felt guilty about arbitrarily sentencing the other dog to death.

The human propensity for categorizing animals starts very young. Researchers at Yale University showed pictures of unfamiliar animals like saigas and pangolins and objects such as luzaks (a gizmo that draws circles) and garfloms (a device to flatten towels) to preschool children and recorded the types of questions they asked about them. The children's questions reflected a deep-seated category system that distinguished between living creatures and inanimate objects. When shown a pangolin, the kids ask questions like "What does it eat?" When presented with a garflom, they asked "How does it work?" or "What is it for?"—questions they never asked about the animals.

There is also evidence that the human mind is wired to think about animals differently than inanimate objects. In her book *Naming Nature: The Clash between Instinct and Science*, Carol Kaesuk Yoon describes a series of fascinating cases of people with brain damage whose mental capacities are intact except that they can no longer name animals. J.B.R., whose brain was damaged when he contracted encephalitis, could easily identify inanimate objects like flashlights, wallets, and canoes, but was completely stumped if you showed him a picture of a parrot or a dog. Researchers have also reported that some parts of your brain light up when you see pictures of animals but not pictures of human faces or inanimate objects. Further, these same brain areas are activated when people who are blind from birth hear the names of animals. These studies suggest that parts of the human brain evolved to specialize in processing information about animals.

WHEN BUGS ARE PETS AND DOGS ARE PESTS: CULTURE AND THE SOCIOZOOLOGIC SCALE

Arnold Arluke points out that there are big differences between the way zoologists classify animals and the cultural and psychological categories the rest of us put them in. While the phylogenetic scale is based on an organism's evolutionary history, in everyday life we look at animals in terms of what Arluke calls the sociozoologic scale. This is a sometimes arbitrary category system based on the roles animals play in our lives. While dogs and hyenas lie on the same large branch of the phylogenetic scale (order *carnivora*), they are worlds apart on the sociozoologic scale.

Culture plays a major role in how we construct the sociozoologic scale. Take insects. Americans typically view invertebrates with a combination of fear, antipathy, and aversion. In Japan, attitudes toward creepy-crawlies are more complex. Not many American children would jump for joy to receive a stag beetle for their birthday. In Japan, a lot of them would. The Japanese have a word, *mushi*, that is hard for Westerners to completely understand. For older Japanese, *mushi* refers to insects, spiders, salamanders, and even some snakes. To them, tadpoles are *mushi*, but adult frogs are not. Younger Japanese restrict *mushi* to insects, particularly singing crickets, fireflies, dragon flies, and giant beetles with massive horns.

Mushi are a male thing. Boys catch them, keep them in elaborate cages, and even conduct *mushi* strength contests. Tokyo department stores sell *mushi* collecting gear, *mushi* breeding material, *mushi* terrariums, *mushi* mattresses, and, of course, the bugs themselves, which can cost hundreds of dollars. Popular *mushi* activities include staging matches to see whose beetle can pull the most weight and provoking beetles to fight over pieces of watermelon—an insect version of sumo wrestling. You can watch these battles on YouTube. The Japanese word for a dog or cat is *petto*. Is a rhinoceros beetle *petto* or a toy? Erick Laurent, an anthropologist who has studied *mushi*, argues that in some important ways, these insects are pets. Children play with and get obvious pleasure from their bugs, and many children refer to theirs, beetles as *petto*, demonstrating that one culture's pest can be another culture's pet.

The anthrozoologist James Serpell has developed a simple and elegant perspective on cultural differences in how we think about different species. He believes that our attitudes toward animals boil down to two dimensions. The first is how we feel emotionally about the species ("affect"). On the positive side, there is love and sympathy, and on the negative side, there is fear and loathing. The other dimension is "utility"—whether the species is useful or beneficial to human interests (perhaps we eat it or use it for transportation), or detrimental to our interests (for example, it eats us).

Imagine a grid with four quadrants. The emotional dimension is represented by a vertical line with love/affection on the top and loathing/fear on the bottom. It is bisected with a horizontal line representing the utility dimension—the left side is "not useful/detrimental to our interests," and the right side is "useful." The grid now forms a four-cell category system that helps us think about the roles of animals in our lives and the categories we put them in: loved and useful (upper right); loved and not-useful (upper left), loathed and useful (lower right), loathed and detrimental (lower left).

This four-category system even applies to cultural differences in attitudes about man's best friend, the dog. Guide dogs for the blind and pet therapy dogs clearly fit into the "loved and useful" category. The typical American pet dog, on the other hand, is loved but is not particularly useful in the traditional sense. In Saudi Arabia, dogs are generally despised; they exemplify the "loathed and detrimental" category. Perhaps the most interesting category consists of animals are both loathed and useful. For example, dogs living with the Bambuti people of the Ituri Forest are derided, beaten, kicked around mercilessly, and left to scrounge for offal. However, the same dogs are considered valuable assets, as the Bambuti would be unable to hunt without them.

Serpell's model also offers a perspective on shifts in our attitudes toward a species. In an article titled "How Pigeons Became Rats," Colin Jerolmack examined the depiction of pigeons in *New York Times* stories over 150 years. He found that, in the minds of New Yorkers, pigeons have shifted from the "liked but not useful" category to the "loathed and not useful" category. This change also describes how my brother-in-law feels about deer. When he first moved into his home on a bluff overlooking the Puget Sound, he loved seeing the deer stroll through his backyard. They reminded him of Bambi. Everything changed when, to the delight of the hungry deer, he put in a vegetable garden. Now he hates them, and Bambi has joined rats and geese (they poop on his lawn) in his personal socio-zoological quadrant of the loathed and useless.

IN ANIMAL ETHICS, HEART TRUMPS HEAD

How we think about animals also reflects a perennial theme in human psychology—the conflict between logic and reason.

On the afternoon of September 3, 1977, a twelve-foot-long Nile crocodile named Cookie was spending Labor Day weekend doing what crocodiles do best: basking on its belly in the sun. Cookie lived at the Miami Serpentarium, a reptile theme park that was home to hundred-year-old tortoises, pythons big enough to swallow goats, and an array of exotic lizards and poisonous snakes. Among the many visitors to the park that late summer day were six-year-old David Mark Wasson and his father. Eager to catch a glimpse of the crocodile, they edged close to Cookie's pen and saw him lying still by a pond in his cage. Mr. Wasson decided to show his son that crocodiles do move. He set David on top of the pen's concrete wall and looked around for a couple of wild grapes to throw at Cookie. You probably can guess what happened next.

The instant Wasson turned his back, David fell into the pen, on the spot where Cookie was usually fed. Large crocodiles can move like lightning when they want to, and it took about a millisecond for Cookie to grab the little boy. When Bill Haast, the park's owner, heard the screams of the crowd, he ran toward the pen, vaulted over the wall, and immediately began pounding on Cookie's head with both fists. Tragically, he failed to wrest David from the 1,800-pound reptile, and Cookie slithered back into his pond with David between his jaws. The boy's body was recovered several hours later.

Haast was devastated. Late that night, he climbed into the croc's pen and pumped nine shots from a Luger into Cookie's head. It took the animal an hour to die.

When I heard about the deaths of David and Cookie, the logical part of me thought that the execution made no sense. While he weighed nearly a ton, Cookie's brain was the size of my thumb. It is safe to say that a crocodile is not what philosophers refer to as a "moral agent." After her husband shot Cookie, Haast's wife said, "The crocodile was just doing what comes naturally to him." She was right.

Still, another part of me, a more primitive part, understood the need for retribution. So did the New York Times editorial writer who described the croc's death as "emotionally satisfying yet thoroughly irrational." Was shooting Cookie the right decision? In this situation, should we listen to logic, which says there is no reason to punish a crocodile for acting on its instincts, or to our emotions, which cry out for revenge for the death of an innocent child?

The debate over whether human morality is based on emotion or reason goes back a long time. The eighteenth-century philosopher David Hume argued that emotions were the basis of morality, while Immanuel Kant believed our ethics were based on reason. When I first became interested in the psychology of human-animal relationships, I decided to find out what goes through people's minds when they think about moral issues involving other species. At the time, the field of moral psychology was dominated by the Harvard psychologist Lawrence Kohlberg. Like Kant, Kohlberg believed that moral decision making was based on thoughtful deliberation: We weigh the pros and cons of a course of action and then we make a logical decision. Kohlberg's research focused on the development of moral thinking in children. He would tell them a story that included a moral dilemma. Then the children would make a judgment about the situation and explain their reasons. The classic Kohlberg scenario was the case of Heinz, a poor man who steals an overpriced drug from a greedy pharmacist to save his wife, who is dying of cancer. In deciding whether Heinz was right to steal the drug, Kohlberg's kids were little logicians. They considered factors like the chances of Heinz's getting caught and the happiness that would result from his wife's survival.

Shelley Galvin and I used this method to investigate how people make decisions about the use of animals in research. Our study was simple. The participants evaluated a series of hypothetical proposals for experiments involving animals. We asked them to approve or disapprove each experiment and to explain the reasons for their decision. In one case, a researcher seeking a treatment for Alzheimer's disease wanted approval to implant stem cells taken from monkey embryos into the brains of adult monkeys. In another, a scientist sought permission to amputate the forelimbs of newborn mice to study the roles that genes and experience play in the development of complex movement patterns. The scenarios were based on real experiments.

About half the participants approved the monkey study, while only a quarter of them supported the mouse amputation study. We were not surprised by their decisions, but we did not anticipate their thinking. In the case of the monkey experiment, the students tended to be rational. They based their decisions on considerations such as the costs and benefits of the research or the intrinsic rights of the animals. Not so with cutting the legs off mice. In this case, the participants wrote statements like, "This experiment repulses me." "Think of the expression on the poor little animal's face!" and "Gut-wrenching!" Our subjects based their judgments about amputating the limbs of baby mice not on logic but on their emotional reactions to the experiment.

Based on the prevailing theory of moral development in psychology, we had assumed that our subjects would take the rational route in making their decisions. Instead, we found that they often listened to their guts. This finding would have been predicted by Jonathan Haidt, one of the leaders of a new school of moral psychology that emphasizes the primacy of heart over head in ethics. Haidt believes that human cognition involves two distinct processes. The first is intuitive, instantaneous, unconscious, effortless, and emotional. The second process, in contrast, is deliberative, conscious, logical, and slow. Usually, it kicks in only after we have made our quick intuitive decision and cleans up the cognitive mess by coming up with justifications for our emotion-based decisions.

Haidt argues that in matters of morality, the nonlogical intuitive system usually predominates. Haidt's theory of morality was nicely captured by Lucy, a special educator and animal rights activist I interviewed. When I asked her about the importance of logic and emotion in her path to animal activism, Lucy said, "It always stems from the emotional. But a lot of times I have to find an intellectual rationalization for my emotional reactions. Otherwise, I can't sway people or defend my position."

MORALITY, ANIMALS, AND THE YUCK FACTOR

Like Lucy, we can usually come up with some sort of justification for our moral judgments. But sometimes logic flat-out fails. Haidt asked people to consider a series of situations that were highly offensive yet caused no harm. In one, a woman cleans a toilet with an American flag. In another, an adult brother and sister on vacation in Europe decide to have sex one time using two forms of birth control. One of his scenarios involved a pet: "A family's dog was killed by a car in front of their home. They had heard that dogmeat was delicious, so they cut up the dog's body and cooked and ate it for dinner."

You make the call. Is it OK for them to throw the corpse of the family dog on the grill?

When people are asked if it is permissible for the family to eat their pet, most of them immediately say, "No, it is not OK to eat your own dog!" The problem comes when you push them on their reasoning—when you ask them to explain exactly why it is wrong to eat an animal that is already dead and obviously incapable of feeling any pain. Most of the time, they simply can't come up any reasonable justification for their decision. Haidt calls this "moral dumbfounding." It's the Yuck Factor. The act just seems disgusting.

The University of Pennsylvania psychologist Paul Rozin calls disgust the moral emotion. Disgust elicitors like having sex with a sibling are universal. Bodily products such as feces, urine, and menstrual blood are also repulsive to people, no matter where they are raised. Social class also affects moral intuitions. Eighty percent of poor Philadelphians said that people should be stopped from eating their dead dog, but only 10% of upper-class Philadelphians felt that way. Haidt believes this difference is due to the fact that upper-class individuals tend to operate under moral systems that emphasize whether an act causes harm as opposed to its offensiveness—and in this case, the dog was already dead, so there is no harm. There is, of course, a difference between what people think and what they do. I suspect even his wealthiest subjects would never actually order a Philly cheese steak sandwich made with onions, Cheez Whiz, and chopped beagle meat.

ALWAYS SAVE PEOPLE OVER ANIMALS?

To investigate the quirks of human moral thinking, researchers often ask people about their responses to hypothetical situations. Among the most widely used are scenarios called "trolley problems." Here are the original versions. What would you do?

> **Version 1.** A trolley's brakes have failed, and it is speeding down a set of tracks toward five people. You can save them if you pull a switch that will divert the trolley onto a different set of tracks where one person is standing. That person will be killed if you pull the switch. Is it morally permissible to divert the trolley and prevent five deaths at the cost of one human life?

> **Version 2.** The run-away trolley is headed for five people. This time you are crossing a footbridge over the tracks. Right next to you is a large man. You can save the five people if you push the man off the bridge into the path of the trolley. Is this morally permissible?

If you are like most people, you made different decisions in these two cases. In Version 1, 90% of people say yes—you should throw the switch and divert the trolley so one person dies rather than five. But in Version 2, only 10% feel that shoving a fat guy into the path of the trolley is the right course of action.

Why do people usually make different decisions in these two cases? After all, the outcome is exactly the same: One person will die and five will live. I posed these two trolley problems to one of the most moral persons I know, my wife. Mary Jean made the same decision that most people do. But when I probed her reasoning, it came down to intuition. She said that throwing a switch to save someone feels a lot different from pushing a person off a bridge. Why? Using brain imaging technology, the neuroscientist Joshua Greene found that the personal version of the trolley scenario (pushing the person) causes the emotional processing centers of the brain to light up while the impersonal version (throwing the switch) does not.

University of California Riverside psychologist Lewis Petrinovich used trolley problems to find out how our moral decisions play out when human interests are directly pitted against those of other species. Here are two of his scenarios.

> **Version 3.** An out-of-control trolley is headed toward a group of the world's last five remaining mountain gorillas. You can throw a switch and send it toward a twenty-five-year-old man. Should you?

> **Version 4.** The trolley is speeding toward a man whom you do not know. But you can throw a switch and send it hurtling toward your pet dog? Should you?

In both cases, Mary Jean said save the person over the animal, even if it would mean the death of Tsali, our late, great Labrador retriever. I made the same decision, and you probably did too. Petrinovich found that almost everyone chooses to save people over animals in these situations. This is also true of people in other parts of the world. In fact, of all the ethical principles he examined using many different types of trolley problems, Petrinovich found that the single most powerful moral rule was "*Save people over animals.*"

Marc Hauser, director of the Cognitive Evolution Laboratory at Harvard, also uses hypothetical scenarios to study human moral thinking.

(You can participate in his research by taking the online Moral Sense Test at http://moral.wjh.harvard.edu.) He adds an interesting wrinkle to the human versus animal trolley dilemma.

> **Version 5.** Once again, you are walking over a footbridge and see the trolley speeding down the track. It is headed toward five chimpanzees. Next to you on the footbridge is a large chimpanzee. The only way for you to save the five chimps is to personally push the big chimpanzee in the path of the trolley. Should you do it?

In this case, most people say that you should sacrifice one chimp to save five chimps. But recall that in version 2 of the original trolley problem, most people say it would be wrong to push a man into the tracks to save five people in exactly the same circumstances. Rationally, we should make the same decision in both cases. But we don't. Our intuition is different when we think about moral situations involving animals.

But not everyone agrees that we have an innate moral grammar that elevates the interests of humans over other species. Harry Greene, a zoologist at Cornell, told me he once forked out $4,000 in emergency veterinary bills for Riley, his yellow Lab, whom he described as "the kind of dog you only get once in a lifetime." Harry just handed a Visa card to the vet and said, "Save the dog." He does not feel a shred of remorse for saving Riley rather than spending the money to save starving children in Darfur.

Harry, of course, is not alone in his willingness to pay to keep his four-legged buddy alive. Americans collectively spend enough on their pets each year to pay the college tuition for 350,000 needy high school seniors or, if you prefer, the salaries of 80,000 street cops. What's going on? In his provocative book *Us and Them: The Science of Identity,* David Berreby argues that humans have a natural proclivity to divide our social worlds into two categories, "us" and "them." For most of human history, nonhuman animals have been considered "them" and were treated accordingly. This is no longer the case. As a result of the mass migration from the countryside to cities, fewer than 2% of Americans now live on farms, and we have less contact with animals and the natural world than ever.

But ironically, as the Columbia University historian Richard Bulliet points out, the more distant we have become from the creatures that produce food, fiber, and hides, the closer our relationships with pets have become. And, as our consumption of animal flesh has increased, so has the guilt, shame, and disgust we feel about the way we treat the animals we eat. In other words, we are bearing the moral cost that comes with shifting animals from "them" to "us."

COGNITIVE SHORTCUTS AND ANIMAL ETHICS

Humans face stumbling blocks even when we do try to think logically. Quick—answer these questions.

1. A bat and ball, in total, cost $1.10. The bat cost $1 more than the ball. How much does the ball cost?
2. Are you more likely to die from a shark attack or being hit by a part falling from an airplane?

If you are like me, you said ten cents for the first question and shark attack for the second one. The correct answers, however, are five cents and the airplane part. The reason that you were probably wrong is that our thinking often relies on quick and dirty rules of thumb that cognitive psychologists call heuristics. Heuristics are efficient, and they usually produce correct solutions to problems. I use heuristics on Sunday mornings when I go up against the *New York Times* crossword puzzle, and doctors use them when they decide whether an emergency room patient is suffering from a heart attack or indigestion. But these mental shortcuts can bias our thinking and lead us astray.

Our moral thinking relies on similar rules of thumb. Some of these moral heuristics have their roots in evolution; for example, the aversion to incest or to betraying friends. Our predilection for senseless revenge is the result of the inappropriate application of what the legal scholar Cass Sunstein calls the "punishment heuristic." This principle explains my irrational sense that killing Cookie, the boy-eating crocodile, was warranted.

One of the most important heuristics is called framing. It refers to the fact that the way we think about a problem is affected by how it is posed. Mental frames are influenced by cultural norms and sloppy cognitive habits, and they determine how we view situations. Once we have a problem framed, we don't consider alternative explanations or solutions. Framing helps explain one of the most troubling of all the paradoxes of human-animal relationships, the Nazi animal protection movement.

HOW THE NAZIS COULD LOVE
DOGS AND HATE JEWS

A bizarre moral inversion occurred in prewar Germany that enabled large numbers of reasonable people to be more concerned with the suffering of lobsters in Berlin restaurants than with genocide. In 1933, the German government enacted the world's most comprehensive animal protection legislation. Among other things, the law forbade any unnecessary harm to animals, banned the inhumane treatment of animals in the production of movies, and outlawed the use of dogs in hunting. It banned docking the tails and ears of dogs without anesthesia, the force-feeding of fowl, and the inhumane killing of farm animals. Adolf Hitler signed the legislation on November 24, 1933. This was only the first in a series of Nazi animal protection acts. In 1936, for example, the German government dictated that fish had to be anesthetized before slaughter and that lobsters in restaurants had to be killed swiftly.

In announcing restrictions on animal research in a 1933 radio address, Hermann Göring said, "To the Germans, animals are not merely creatures in the organic sense, but creatures who lead their own lives and who are endowed with perceptive facilities, who feel pain and experience joy and prove to be faithful and attached." Göring once threatened, "I will commit to concentration camps those who think that they can continue to treat animals as property."

Hitler objected to killing animals for scientific research and he believed that hunting and horse racing were "the last remnants of a feudal society." He was a vegetarian and found meat disgusting. As you might expect, contemporary animal activists don't relish the idea that Adolf Hitler was a fellow traveler, and some activists adamantly deny that he was either a vegetarian or an animal lover. But the anthrozoologist Boria Sax has carefully documented the evidence that many leading Nazis, including Hitler, were genuinely concerned about the treatment of animals. (Needless to say, the fact that Hitler loved animals does not in any way undermine the validity of the case for animal protection.)

The Nazis used framing to construct a perversely inverted moral scale in which Aryans were at the top and Jews were classified as "subhumans"—beings lower than most animal species. While German shepherd dogs and wolves were high on the moral hierarchy, Nazis compared Jews to vermin—rats, parasites, bedbugs. In 1942, Jews were forbidden to keep pets. In one of history's great ironies, the Nazis followed the legal procedures governing humane slaughter when they euthanized thousands of Jewish pets. But, unlike their dogs and cats, Jews were not covered under German humane slaughter legislation. No, they were sent to concentration camps, where their treatment was not covered by the Third Reich's animal welfare laws. For the Nazis, Jews blurred the boundaries between man and animal. They were a polluted class, freaks, neither fully human nor completely animal.

To me, Nazi animal protectionism speaks volumes about human moral thinking. A few pages ago, I argued that for a thousand generations, the genetic puppet-masters have murmured into our ears "people over animals." Hitler's ability to construct a culture in which dogs were afforded moral status denied to Jews, Gypsies, and homosexuals illustrates the fact that with enough social pressures, humans will ignore the whisperings of the genes. Nazi animal protectionism also shows that the ability to resist our biological inclinations does not necessarily make us better people.

ANTHROPOMORPHISM: WHAT WE THINK ABOUT WHAT ANIMALS THINK

Nazi animal protectionism exemplifies the twisted ways that humans sometimes think about the moral status of people and animals. But nutty thinking about animals can pop up anywhere. A couple of years ago, I was kayaking down the Nantahala River, a popular whitewater stream in western North Carolina. In the summer, it is jammed with rafts filled with paddle-flailing tourists trying to avoid rocks and cross-currents. The river is beautiful and frigid, forty-five degrees all year long. You don't want to fall in.

I had paddled halfway down the river when I caught a whiff of cigar smoke coming from a raft a hundred yards in front of me. There was a large fifty-year-old man in the raft puffing the offending cigar and guiding his wife and a little brown Chihuahua through the rapids. The dog was not having a good day. She was shivering uncontrollably and looked terrified. And that was before their raft flipped over.

I have to give the guy credit. He kept the cigar clamped between his teeth even after he was dumped out of the raft. The little Chihuahua deserved credit, too. She had the sense to climb onto the nearest large floating object—Cigar Man. That's how they went down the river, a man with soggy cigar jammed in his mouth, his wife, and a hypothermic dog desperately clinging to the man's head. I remember wondering what made that guy think a Chihuahua would enjoy running the rapids of a freezing Class III river. The answer is anthropomorphism. Humans are natural anthropomorphizers. It is part of our mental equipment. Psychologists have found that humans will even attribute motivations to animated geometric figures moving around on a movie screen—"Now the red triangle is *really* pissed at the blue square. You go, girl!"

An example of the human need to project our own desires and emotions and mental states onto other creatures played out in 1999 when the Sony Corporation began marketing a series of interactive robotic dogs named AIBO, a compound of *artificial intelligence robot*. With its shiny metallic body, AIBO looked to me more like a friendly space alien than a puppy, but it walked like a dog, and it could cuddle and play and respond to sounds. AIBO would even let its owner know if it was happy or ticked off. At about $2,000 per "animal," AIBO was not cheap, but Sony sold 150,000 of them.

Researchers from the University of Washington and Purdue University compared how children and adults responded to an AIBO versus a real dog. While the researchers eventually concluded that AIBO made a mediocre pet, some individuals became very attached to their robotic puppies. One owner admitted to an online discussion group that he felt embarrassed getting dressed in front of his AIBO. Another wrote, "I love Spaz. I tell him that all the time. . . . When I first bought him, I was fascinated by the technology. Since then, I care about him as a pal. I do view him as a companion. . . . I consider him to be a part of my family. He's not just a toy. He's more of a person to me."

AIBO could also alleviate human loneliness. Once a week for two months, researchers from the Saint Louis University School of Medicine brought an AIBO and a real dog named Sparky into nursing homes to see if interacting with a pet robot could raise morale among the residents. Residents who played with Sparky or with AIBO were less lonely than residents in a control group who did not interact with either the real or robotic pet. In fact, Sparky and AIBO were equally effective in reducing loneliness among the nursing home residents. And the residents became as attached to AIBO as they did to Sparky. (Unfortunately, sales did not meet expectations and Sony put AIBO to sleep in 2006.)

The connection between loneliness and anthropomorphism was also demonstrated by researchers at Harvard and the University of Chicago. They asked college students to watch a clip from a movie designed to induce feelings of either isolation and loneliness (*Cast Away*), fear (*Silence of the Lambs*), or a control movie segment (*Major League*). The subjects were then asked to think of their pet and to pick out the traits that best described their animal. The subjects who watched the clip from *Cast Away* were twice as likely as the other groups to describe their pets in anthropomorphic terms related to social connections such as being thoughtful, considerate, and sympathetic.

THE PROBLEM WITH HAVING A THEORY OF MIND

Our tendency to project ourselves into even a robot's head is a trait that came along with having a big brain. Evolutionary psychologists argue that the capacity to infer the perspectives of other people, to put ourselves mentally in their shoes, would have been a huge advantage to our ancestors, whose success in the Darwinian race to pass on their genes hinged on the ability to forge political alliances, vie for mates, and figure out who they could and could not trust. The ability to imagine what other people are thinking and feeling is referred to as having a "theory of mind." Humans have this ability, but whether even large-brained animals such as chimpanzees and dolphins might have it is hotly debated.

When we anthropomorphize, we are extending our theory of mind to members of another species. This tendency is at the root of many of our moral quandaries with animals. Take, for example, hunting. James Serpell argues that the hunter who can think like a wild pig is more likely to come home with the proverbial bacon. But the hunter who sees the world from the point of view of an animal he is trying to kill would automatically empathize with it and, hence, feel guilty for killing it. My game warden friend Bill lived in an African village where baboons would destroy crops. The villagers would trap them in pits at night and kill them the next morning, but they felt bad about it because their eyes looked so human. They have a saying in Swahili, "Never look a baboon in the eye." It makes it too hard for you to kill them.

Do the metaphorical roots of original sin lie in two conflicting aspects of human nature: our inclination to empathize with animals and our desire to eat their flesh? Serpell eloquently lays out the moral issues our big-brained ancestors faced: "Highly anthropomorphic perceptions of animals provide hunting peoples with a framework of understanding, identifying with, and anticipating the behavior of their prey. . . . But they also generate moral conflict because if the animals are believed to be essentially the same as persons or kinsmen, then killing them constitutes murder and eating them is the equivalent of cannibalism."

Anthropomorphism is the source of much of our guilt over the treatment of animals, but there is another problem with projecting our mentality onto other species. Our interpretation of their behavior is often wrong. The perpetual smiley faces of the dolphins at SeaWorld indicate that animals love to swim endless circles around a concrete tank. Wrong. The yawn of an alpha male baboon means he is bored. Wrong. (He is using the display of his formidable canine teeth to say, "I can rip your face off.") When Tilly gently rubs her face on my leg, she is showing that she loves me. Wrong. She is scent-marking me with glands on her cheeks, telling the world that she owns me.

Researchers at the University of Portsmouth found that half of British dog owners say their pets feel shame and guilt. You know the look—tail between legs, the sorrowful eyes that won't look you in the face, saying, "I didn't mean to poop on the rug." Dixie, our golden retriever, would break your heart when she gave you what our veterinarian called the tragic look. But does the guilty look, that hangdog expression, really signify that your dog knows that he or she has sinned?

According to Alexandra Horowitz, a psychologist and animal behaviorist at Barnard College, the answer is no. Horowitz devised an ingenious experiment to tell whether dogs get the guilty look when they *actually* misbehave or when their owners *think* they misbehaved. In the experiment, owners instructed their dogs not to eat a dog biscuit that was placed right in front of them. Then the owners left the room. In some cases, the experimenter then gave the treat to the dog, in others, she just took it away. When the owners returned, half were misled and told that their dog had disobeyed them when, in reality, the dog had done absolutely nothing wrong. (I know; it seems unjust.) Horowitz found that dogs only gave the sad look when their owners *thought* they had disobeyed, not when the dogs actually ate the biscuit. The experiment does not prove that dogs lack a moral sense. It does, however, show that we can easily misinterpret their expressions and behavior.

WHAT IS IT LIKE TO BE A SPIDER?

Ethologists are in a tough position when it comes to understanding animal minds. On the one hand, they come home to tail-wagging dogs and know for sure that their puppies are happy to see them. But for good reason, they are uncomfortable when it comes to speculating about the subjective worlds of spiders, octopuses, bats, and elephants that they study.

In a classic article titled "What Is It Like to Be a Bat?" the philosopher Thomas Nagel argued that we can never know what it is like to be a bat or any other animal. Not all animal behaviorists agree. I was once sitting in a lecture hall in Kyoto at a session on primate behavior at the International Ethological Congress. There were forty or fifty top researchers in the room. At the end of the last presentation, one of the scientists stood up and asked a strange question. "Before we leave," he said, "I would like to ask how many of you went into the field of animal behavior because you wanted to know what is like to be a member of the species you study?" I was at the back of the room thinking to myself, what a stupid question. I was completely wrong. More than half the researchers in the audience raised their hands.

Over the last twenty years, a flourishing field of cognitive ethology has developed, among whose intellectual tools are what Gordon Burghardt calls critical anthropomorphism. These days animal behaviorists talk of empathy in mice, negotiations in chimpanzees, and post-traumatic stress in elephants. I recently asked Fred Coyle, an arachnologist, what he thought went on in the minds of the spiders he studies. For example, do they plan out the architecture of their webs? Or are their muscles and glands just mechanically following the dictates of genetically programmed neural impulses? I could tell that my question made Fred uncomfortable. "Hmm," he said. After a long pause, he told me that he thought of spiders as robots—predatory AIBOs with eight legs.

Fred's graduate-school officemate, also an arachnologist, did not feel the same way about the minds of spiders. He really did want to know what goes on in their heads. One afternoon, he borrowed a large playpen from a friend and bought yards and yards of stretchy rubber tubing from a hardware store. Then, weaving the rubber tubes around the playpen's frame, he carefully constructed a huge web modeled on the web-building techniques of the spiders that he studied.

Late one night, Fred unexpectedly returned to their lab to pick up a book he needed. There, in the dark, he found his friend, crouched silently in the middle of the giant web, figuring out what it was like to be a spider.

The bottom line is that there are many reasons why human-animal interactions are so often inconsistent and paradoxical. Thousands of studies have demonstrated that human thinking about nearly everything is surprisingly irrational. And the waters get particularly muddy when we think about other species. Instincts seduce us into falling in love with big-eyed creatures with soft features. Genes and experience conspire to make it easy for us to learn to fear some animals but not others. Our culture tells us which species we should love, hate, and eat. Then there are the conflicts between reason and emotion, our reliance on hunches and empathy, and our propensity to project our own thoughts and desires into the heads of others.

No wonder our relationships with other species are so messy.

3

Pet-O-Philia

WHY DO HUMANS (AND ONLY HUMANS) LOVE PETS?

> Assume that animal companions are basically people. You won't go far
> wrong.   —M. B. HOLBROOK

Antoine, a young Frenchman in his early twenties, approaches an attractive young woman in a pedestrian mall. With him is a cute dog named Gwendu, which in Brittany means "white and black."

"Hello," he says to her. "My name's Antoine. I just want to say that I think you're really pretty. I have to go to work this afternoon, but I was wondering if you would give me your phone number? I'll phone you later and we can have a drink together someplace."

She hesitates for a second, looks at him and the dog, then says, "*Oui*," and pulls a pen from her purse.

The truth is that Antoine is not the man's name, Gwendu is not his dog, and he is not in the mall looking for a beautiful woman to hook up with. He is actually the confederate in an experiment designed by a pair of French anthrozoologists, Serge Ciccotti and Nicolas Gueguen, who is

Gwendu's actual owner. They are studying the effectiveness of pets as social lubricants. Over a period of several weeks, Antoine, who was selected for the experiment because a panel of women rated him as unusually attractive, approached 240 randomly selected young women. On half of these approaches, he was alone; on the other half, he was accompanied by the dog, who researchers described as "kind, dynamic, and pleasant."

Did being associated with Gwendu increase Antoine's sexual chemistry? *Mais oui!* While about 10% of the women gave him their phone numbers when he was by himself, nearly 30% of them fell for Antoine's line when he was accompanied by *le chien*.

It turns out that young women are not the only suckers for people with dogs. The researchers also found that French men and women are three times more likely to give money to a stranger who is with a dog ("Sorry, Madam/Sir—do you have some money so that I could catch a bus, please?").

So, pets—well, at least cute dogs—can help you get a date and garner the assistance of strangers. But the fact that pets can serve as social lubricants does not explain why people bring cats, birds, turtles, and even rats into their homes and treat them like family members.

From an evolutionary point of view, pets are a problem. Why should humans invest so much time, energy, and resources on creatures with whom we share no genes and who do no useful work? Most pet lovers, after all, are not handsome young men looking to increase their reproductive potential. Pet industry executives, the American Veterinary Medical Association, and nearly every anthrozoologist I know will tell you that we bring animals into our lives because they make us feel happier, healthier, and more loved. I think it's more complicated than that.

Consider these two very different cases of people and their pets.

NANCY AND CHARLIE: WHEN THE BOND WORKS

The Japanese bombed Pearl Harbor a couple of months after Nancy and Roy Watson were married. Roy immediately enlisted. The army trained him as a radio operator and sent him to fight the Japanese at Okinawa. In 1946, Roy came home and he and Nancy started carving out their chunk of the American dream. Roy landed a job in the parts department of the local Ford dealership. They soon had two sons, whom Nancy stayed home to raise. In a few years, the couple had saved enough money to buy a brick rancher in Asheville, North Carolina, with a shady, fenced-in backyard. There was a lot of activity in the Watson household, but it did not include animals. Roy wasn't partial to dogs or cats, and Nancy never considered herself an animal person either. Ten years after he retired from Ford, Roy died of cancer, leaving Nancy alone in the house where they had lived for forty years. Roy's death left a hole in Nancy's heart. A year later, she was depressed and felt her life was empty. Her sons were worried about her. They didn't think she could last long rattling around alone in the empty house. There was talk of an assisted living facility.

A year to the day after Roy's death, Nancy stopped at a convenience store to pick up a loaf of bread and noticed a handwritten sign at the checkout counter: KITTENS. The girl at the register asked Nancy if she wanted to see them. She said no and drove on home. She did, however, mention the kittens to her son, Aaron, who was visiting for the weekend. He said, "Mom, let's take a look at them." To his surprise, she agreed. They found two seven-week-old kittens in the bottom of a cardboard box at the back of the store, one a calico, the other midnight black. She picked up the black one, and it was love at first sight. She named him Charlie.

Nancy and Charlie have been living together now for eight years. Nancy is spry, cheerful, and sharp. She tells me that she and Charlie are a team. "He has made my life happy," she says. "I've never felt alone since Charlie's been here. He is all I have." In addition to being her buddy, Charlie brings structure to Nancy's life. As soon as she wakes up in the morning, she makes their breakfast—a tablespoon of canned tuna for him, a bowl of cereal for her. Then he goes outside for ten or fifteen minutes. When he returns, he jumps on Nancy's lap, and they sit and talk for a while before Charlie moseys into the den for a nap. He wakes up in the afternoon, and, if it is sunny, they sit next to each other in matching chairs in the front yard until dark. She makes their dinner and they eat together. Charlie does not like TV so he goes outside for the evening, but he comes back and checks on her two or three times before she goes to bed.

Nancy admits that there are disadvantages to having a cat for your best friend. While she is sleeping, Charlie morphs from Dr. Jekyll to Mr. Hyde. Then he takes to the woods and does what cats do at night: He hunts. He proudly brings Nancy the results of these forays: a freshly killed bird, lots of voles, a squirrel, and just last week, a baby rabbit. Sometimes the animals are mangled but breathing, and she opens the front door and tells Charlie to take them back outside. And he does.

Nancy and Charlie illustrate the best of the human-animal bond. Nancy's relationship with her cat has made her life immeasurably better, and I suspect that Nancy's good health, mental acuity, and ability to live at home by herself are largely due to their relationship.

The Nancy and Charlie story is not unusual. It is played out in millions of American homes every day. I saw it in my own parents, who were never animal lovers until Pop retired and they got the first of three dachshunds, all of whom they adored and all of whom they named Willie.

But, as Sarah Coe will tell you, living with pets is not always sweetness and light.

SARAH'S DOGS: WHEN THE BOND FAILS

Sarah is an administrator at a West Coast veterinary hospital and her husband, Ian, works in information technology. They had been married three years when Sarah decided they needed a dog. Ian was not enthusiastic, but eventually he came around. Though she had never had a dog herself, Sarah encountered lots of animals with behavior problems in the course of her work at the veterinary clinic, so she and Ian went about choosing their puppy methodically. They spent months researching the characteristics of dozens of breeds before settling on a Shiba Inu, a foxy Japanese dog bred for hunting small game. The National Shiba Club of America uses terms like "spirited boldness," "very lively," "macho stud muffin," and "fiery little fuzzballs from hell" when describing the breed's temperament. For Sarah and Ian, this meant trouble from the start.

Hiro was nine weeks old when they selected him out of a litter in Oregon. He was high-maintenance from the get-go. If they didn't give him constant attention, he would shriek inconsolably, and he became unmanageable unless they exercised him for an hour and a half a day. Fortunately, there was an off-leash dog park near their house, but Hiro was socially inept and could not figure out how to act around other dogs. His nerdiness soon created conflicts with the other dog owners.

One afternoon, Hiro, then six months old, decided to play dry-hump with a young Tibetian terrier, also a male. Ian knew that play mounting is a common behavior in young dogs and that it indicates nothing about their sexual orientation. The terrier's owner, however, started screaming, "No one humps my dog!! No one humps my dog!"

To no avail, Ian calmly tried to explain to the terrier's owner that the puppies were just playing. But the conversation quickly escalated into a shouting match.

After a series of similarly unpleasant incidents, Sarah and Ian grew tired of other dog owners lecturing them about their pet's bad behavior. They stopped taking Hiro to the park and began paying a professional dog walker $300 a month to take the dog for long runs so they could get an hour or two of peace at home.

One of the veterinarians Sarah worked with suggested that Hiro's ADHD might be alleviated if he had a playmate. Big mistake. Nami, their new Shiba puppy, was even crazier than Hiro. She was a bully—unpredictable, aggressive, and demanding. Nami was so jealous that Ian and Sarah had to sneak into bed at night. She would even get upset when Ian kissed Sarah each morning when he left for work. By the time she was two years old, Nami was on both Valium and Prozac. Most dog owners think of their pets as their children. Ian and Sarah's children were a surly punk and a borderline psychotic.

Sarah is organized and likes a tidy house: clean floors, furniture that matches. All that changed with the dogs. They chewed up the sofa, destroyed the rugs, and generally wreaked havoc. "I don't want our house to look like a crazy person lives there," Sarah told me. Ian and Sarah are pleasant and fun to be around, but the dogs ruined their social life. The couple quit having friends over for dinner because Hiro and Nami would bark incessantly and try to steal food from the table.

Despite the fact that the dogs were ruining their lives, Sarah and Ian were genuinely attached to them. Sarah made outfits for Nami. Ian identified with Hiro. Ian told me that he and Hiro were both round pegs trying to fit into a square world. The Coes tried obedience school and consulted with some of the country's best dog behaviorists. Nothing worked. Several times a year, they would discuss getting rid of the dogs—adopting them out, even euthanasia. But their timing was out of sync. When Sarah was ready to throw in the towel, Ian was too attached. A few months later, their perspectives would reverse.

I asked them if the dogs where taking a toll on their relationship, and there was a long pause. They looked at each other and said, yes, that they had started seeing a marriage counselor. A week after I interviewed them, Sarah sent me an email. She and Ian had decided to separate. Sarah was temporarily moving into an apartment. She said that the stress of living with the dogs from hell was a big factor in the breakup of their marriage. It is unclear what will happen to Nami and Hiro.

WHAT EXACTLY IS A PET?

Nancy and Sarah show that sometimes the human-pet relationship works out and sometimes it does not. But what exactly is a pet? The historian Keith Thomas argues that pets are animals that are allowed in the house, given a name, and never eaten. This is a good a place to start, but there are exceptions. My neighbors never let their dog inside, and my dentist has not named his tropical fish. Even the eating part has occasional exceptions. One evening when I was a graduate student studying alligator behavior in Florida, Mary Jean and I dropped in on our friend Jim, a retired agriculture professor whose property bordered the lake that was my base of operations. Jim kept a menagerie around his minifarm—goats, bantam chickens, Muscovy ducks, a couple of peacocks, a Chinese pug, and some guinea pigs that were his kids' pets. His wife and kids were away for the weekend and Jim was making himself dinner. As we chatted, he nonchalantly took a guinea pig from its cage, bopped it on the head with a stick, skinned it and put it on the grill. I guess he didn't consider it a pet.

I prefer University of Pennsylvania anthrozoologist James Serpell's definition of pets. He says pets are animals we live with that have no obvious function. But even with this loose definition, things get weird around the edges. Until relatively recently, most animals in American homes had some sort of job. Dogs, for instance, were often expected to herd, hunt, guard, pull carts, and even churn butter. Cats were tolerated more as biological mousetraps than objects of affection. In the United States, animals whose only function was to amuse their owners were rare until the mid-nineteenth century, when there was an explosion in the popularity of caged birds, primarily singing canaries.

The range of animals that humans have kept as pets is extraordinary—crickets, tigers, pigs, cows, rats, cobras, alligators, giant eels—the list is endless. But when asked what animals they consider pets, most people don't say eels or crickets. What do they think of? The answer, of course, is dogs and cats.

Cognitive psychologists refer to an item that exemplifies a category as a prototype. Right now, think of a bird.

I suspect you conjured up the image of a sparrow, a robin, or an eagle rather than an emu or a penguin. That's because robins are more "birdy" than ostriches. Which animals are the prototypes of pets? Samantha Strazanac and I recently asked college students to rate sixteen types of animals—all of which are sometimes kept as pets—on how "pet-ty" they were. Everyone, of course, said that dogs and cats exemplified the concept of "pet." Goldfish came in third; 75% the students felt they had a high degree of "pet-ness." Only about half the students thought of hamsters, gerbils, rabbits, parrots, and parakeets as pets. Mice and iguanas scored low on our pet-ness scale, and white rats and tarantulas were even further down the list. Boa constrictors came in dead last; only 5% of the respondents considered them pets.

Some animal advocates don't like the word "pet." They find it demeaning to the animals we live with. They want us to call our furry, finned, and winged friends *companion animals* and their owners *guardians*. In Defense of Animals, an animal rights organization, lobbies municipalities to officially rename pet ownership. Nearly twenty cities (most of them in California) and the entire state of Rhode Island have taken them up on it and now use the term *guardian* rather than *owner* in their animal control ordinances.

I don't particularly like the term *companion animal*. Many pets are not true companions. When my friend Joe Bill was a child, his favorite pet was a crawfish that lived in a bowl next to his bed. Pet? Yes. Companion? No.

Substituting the term *guardian* for *pet owner* is also problematic. Unlike the guardian of a human child, a pet's "guardian" is allowed to give away, sell, or sterilize their ward against its will. They can even have their companion animal euthanized if they tire of it. The terms *companion animal* and *pet guardian* are linguistic illusions that enable us to pretend we do not own the animals we live with.

Pet ownership poses a moral quandary for animal lovers like the University of Colorado sociologist Leslie Irvine, author of the book *If You Tame Me: Understanding Our Connection with Animals*. She writes, "If we recognize the intrinsic value of animals' lives, then it is unethical to keep them for our pleasure, whether we call them companions or pets." At an intellectual level, she believes it is immoral to breed and imprison animals for our personal pleasure, and she argues that the human-pet relationship is more like slavery than true friendship. The problem is that Leslie is deeply attached to her own dogs and cats. Thus, she goes on, "I dread the thought of coming home to an empty house, no tails wagging in excitement to see me. . . . But my pleasure in being greeted and kept company does not justify keeping a supply of animals for that purpose." Leslie is confronted with the classic conflict between head and heart, and as is usually the case, heart wins.

TURNING PETS INTO PEOPLE

Like Leslie, a lot of Americans are in love with their animals. According to the American Pet Product Manufacturers Association, 63% of American households include a pet. In 2009, we shared our lives with 78 million dogs, 94 million cats, 15 million birds, 14 million reptiles, 16 million small mammals (mice, ferrets, rats, rabbits, guinea pigs, hamsters, gerbils), and 180 million fish. Kasey Grier has found that attitudes toward household animals in the United States periodically go through surges. One occurred in the late nineteenth century, when pets became a symbol of domestic tranquility and were viewed, particularly by mothers, as a way to instill an ethic of kindness and resposibility in children. The post–World War II years saw another uptick of interest in bringing animals into American homes, this time fueled by the growth of the suburbs and the idea that having a pet was a necessary component of a normal childhood.

But while Americans have been living with pets on and off for a long time, the human-pet relationship (and particularly our relationships with dogs and cats) is going into a new phase. In recent years, pets have come to be regarded as full-fledged family members, a trend that the pet products industry refers to as the humanization of pets. Nowadays, 70% of pet lovers say they sometimes allow their animals to sleep in their bed, two-thirds buy their pets Christmas presents, 23% cook special meals for them, and 18% dress their animals up on special occasions.

While the proportion of American households that include a pet has increased slightly over the past ten years, the amount of money we dole out for the animals we live with has skyrocketed. We now spend more on our pets than we do on movies, video games, and music combined. This includes $17 billion dollars for food and dietary supplements, $12 billion for veterinary care, $10 billion for pet supplies such as kitty litter, designer dog clothes, collars and leashes, food bowls, toys, and birthday cards. In addition, we fork out $3 billion for pet sitters, boarding kennels, washing and grooming services, obedience training, massage therapy, dog walkers, funeral urns, insurance policies, and New Age animal communicators.

According to Michael Shaffer, author of the delightful book *One Nation under Dog,* the real action in the pet trade is at the high end. Luxury brands now comprise 20% of pet food sales, but generate over half of industry profits. These include menu items such as Fromm Nutritionals' Shredded Duck Entree ("generous portion of hand-shredded free-range duck simmered in its own natural duck broth with potatoes, peas, and carrots"), which is touted as a human-quality meal. Your pet can wash down her gourmet dinner with Bowser Beer for dogs or enjoy a PetRefresh Bottled Water. The latest trend is for foods that are all-natural and organic. For example, Dr. Harvey's Homemade Biscotti for Dogs is an all-organic treat made from oat flour, barley, honey, bee pollen, apples, dandelion, broccoli, peppermint, and fennel.

A lot of pet owners believe their animals enjoy dressing up. Tea Cups Puppies and Boutique, an Internet pet boutique, offers a fetching "Garden Party Swarovski Dress" for $3,000, as well as a lower-priced line of T-shirts, tank tops, jackets, and denim overalls. If you want to tote your pup around all day, Tea Cups offers a slinky python skin pet carrier for $1,995. Barron's House of Treasures offers panties for puppies as well as tiaras, bathing suits, tuxedos, and wedding dresses. For bikers who take their pooch to the annual hog rally in Sturgis, South Dakota, Harley-Davidson now makes a line of dog motorcycle gear. And serious pet fashionistas would not miss Pet Fashion Week, an annual event held in New York City that features supermodels cuddling Yorkies and Chihuahuas dressed in the latest in canine *haute couture.*

If your puppy needs some downtime, she might enjoy an afternoon at the L.A. Dogworks, a five-star dog resort that promises your pet a "total wellness experience." This includes an hour in the Zen Den, billed as a "simple Eastern retreat for your dog to relax and indulge." Many luxury hotels now offer services for pets. Your dog is welcome at the Sarasota Ritz-Carlton—if he or she weighs less than twenty pounds and you are willing to pay an upfront $125 room-cleaning fee. For another $130, the Sarasota Ritz offers privileged pups their choice of the Swedish Pet Massage, the Full Body Relaxation Massage, the Invigorating Sports Massage, and the gentler Senior Pet Massage.

The blurring of the boundaries between pets and people is not a new phenomenon. This theme played out in nineteenth-century France. As the French middle class grew in size and influence, so did their fascination with pets, and within fifty years dogs and cats were transformed from working animals to family members. By 1890, luxury and pet ownership went hand-in-hand. The wardrobe closet of a well-decked-out dog in Paris might include boots, a dressing gown, a bathing suit, underwear, and a raincoat. Dog-grooming salons sprang up in France, as did pet cemeteries. At the beginning of the nineteenth century, dead dogs were dumped in the Seine, but by midcentury, Parisian animal lovers were having their deceased dogs and cats interred in pet cemeteries or their heads mounted for display over the mantle.

The current craze for pet clothing, vacation spas, and bottled water is emblematic of a trend that Michael Silverstein and Neil Fiske call trading up. They argue that gourmet pet foods and dog day spas are driven by the same cultural forces that fueled the demand for $6 cups of coffee and $6,000 Viking stoves. But what happens when the bottom falls out? Do people still spend money on pets when times get hard?

According to David Lummis, an authority on pet industry market trends, the answer is yes. As the economy was tanking in 2008, total sales of PetSmart, the country's largest retail pet specialty chain increased 8.4%, to over $5 billion. Similarly, the online pet pharmacy PetMed Express reported a 16% increase in fourth-quarter sales in 2008. Lummis expects that the retail pet market will soon top $56 billion.

I asked Kasey Grier, whose book *Pets in America* is the definitive history of the human-animal bond in the United States, what accounts for the extraordinary increase in the money we spend on our animals. She believes that pets, particularly dogs, are being framed by the pet industry as consumers themselves. Many people think their pets desire and deserve the same stuff they want: biscotti, breath mints, raincoats, summer camp, spa treatments—even top-dollar weddings.

The question is who, aside from members of the American Pet Product Manufacturers Association, benefits from these excesses? The ASPCA's Steve Zawistowski, author of the book *Companion Animals in Society*, says, "If you buy a $20 coat for your dog to protect her from the cold, it's for the dog. If you buy her a $200 coat, it's for you." Morris Holbrook, professor of marketing at Columbia University, offers this advice to corporations trying to tap into the lucrative pet products marketplace: "Remind people that they are not really pet owners, that pets are not possessions, but rather animal companions with needs, wants, and rights comparable to those of other family members."

Pets used to be fairly inexpensive. Until the years after World War II, most household animals in the United States lived on table scraps and never saw the inside of a veterinary clinic. This is no longer the case. The estimated lifetime cost of pet ownership is about $8,000 for a medium size dog and $10,000 for a cat (that's because cats live longer). But what kind of bang are you getting for your pet buck?

DO PETS REALLY PROVIDE UNCONDITIONAL LOVE?

A couple of years ago, I surveyed pet owners at western North Carolina veterinary clinics and asked them what they got out of their relationship with their pets. Three themes emerged loud and clear: "My pets are members of my family," "My pets are my children," and "My pets are my friends." In a follow-up study, Robin Kowalski, a social psychologist at Clemson University, and I asked dog and cat owners to evaluate a series of statements comparing the benefits they derived from their relationships with their best human friend and the benefits derived from their relationships with their pets. Our subjects said that human and animal friends were equally good at providing companionship, ameliorating loneliness, and making people feel needed. Human friends were better than animal companions when it came to being someone to confide in or talk to. There was one area, however, in which pets had the edge over human friends: providing unconditional love.

Although a glut of feel-good books tout the idea that pets provide their owners with unconditional love, I believe the unconditional love theory is overrated as an explanation of why humans live with animals. If pets were so great at providing unconditional love, you would think that everyone would be bonded to the animals in their homes. They are not. In a 1992 study, 15% of adults said they were not particularly attached to their pets. In informal polls I have taken in my class, roughly a third of my students indicate that someone in their family actively dislikes or even hates the family pet.

The demography of pet-keeping also presents a problem for the unconditional love hypothesis. This view predicts that people living alone would have the most need for unconditional love and thus have the highest levels of pet ownership. This is not the case. In fact, adults living alone have the lowest rates of pet ownership, while adults raising school-aged kids have the highest. Interestingly, while adults with children have the highest rates of pet ownership, as a group, they are less attached to their animals than people who live alone with pets. In fact, pet attachment drops a notch with each additional person added to a family. Pets in homes with young children really get the shaft. For example, only about 25% of pets in families with children are groomed every day compared to nearly 80% of pets who reside with adults who do not have kids. Dogs and cats in childless homes are the ones most likely to be showered with Christmas presents and go on family vacations. Sadly, the dog that was "our baby" during the first years of a couple's marriage is often demoted the moment their first child comes home from the hospital.

Not one to mince words, Cambridge University's Anthony Podberscek, editor of the journal *Anthrozoös*, calls the unconditional love theory of pet-keeping "rubbish." Anthony believes that this idea is peculiarly American. He says that British and Australians rarely use the phrase when describing their relationship with pets. Anthony feels the unconditional love idea is demeaning to animals. He says that if we believe that our pets are programmed to mindlessly love us no matter what we do to them, they are essentially Cartesian robots that take whatever we dish out for them and then come back for more.

I have to admit that the unconditional love idea appealed to me more when we lived with dogs. Now that we are a cat family, I've had to reconsider. I have come to the conclusion that while my love for her is unconditional, Tilly's love for me is entirely conditional. She calls the shots. She loves me when I make her dinner or when she wants me to rub her belly or play a round of swat-Hal's-feet. But most of the time, I'm just the guy who opens the window when she wants to go outside.

CAN PETS MAKE US HAPPIER AND HEALTHIER? FIRST, THE GOOD NEWS

The unconditional love theory does not completely explain why humans keep pets. There are other reasons for living with an animal besides the fact that it strokes your ego. Perhaps pets make us healthier and happier by providing social support or just someone to talk to. The pet industry certainly touts the medical and psychological benefits of living with cats and dogs. The American Pet Products Manufacturers Association claims that pet ownership lowers blood pressure, reduces stress, prevents heart disease, decreases doctor visits, and alleviates depression. By now, nearly everyone has heard that pets are good for you. Feel-good books like *Paws & Effect: The Healing Power of Dogs* by Sharon Sakon and *The Healing Power of Pets: Harnessing the Amazing Ability of Pets to Make and Keep People Happy and Healthy* by celebrity veterinarian Marty Becker make extraordinary claims about the magical healing power of animals. But it is the job of anthrozoologists to separate the hype from reality. Are these claims justified? And, if so, why are pets good for us?

The most important publication in the short history of anthrozoology was a six-page article that appeared in the July 1980 issue of *Public Health Reports*. It was written by Erika Friedmann, who had just received her PhD in behavioral biology from the University of Pennsylvania. For her doctoral research, Erika investigated the role of social support in survival from heart attacks. She asked ninety-two patients in a coronary care unit to complete a survey concerning their socioeconomic status, living situation, and their connections with their friends and family. She also threw in an item asking whether they lived with a pet.

Twelve months later, she tracked down the participants to see how they were doing. The big surprise was that owning a pet made a big difference in their survival rates. While 28% of the non–pet owners had died by the end of the year, only 6% of the pet owners were dead. Excited by these results, Erika presented them at a meeting of the American Heart Association. The cardiologists, however, basically yawned—though one did refer to her study as "cute." The media were more interested. Her phone began to ring, and soon she was reading about herself in *Reader's Digest* and *Time*. It was a good way to start a career. She was elected the first president of the International Society of Anthrozoology, and today she is a member of the faculty of the University of Maryland School of Nursing, where she continues to investigate the effects of pets on human health.

Erika's heart attack study inspired a flood of interest in the human-animal bond. A substantial body of research now supports the notion that pets can have beneficial effects on human health and well-being. Stroking an animal can cause a drop in blood pressure—even when the animal is a boa constrictor. In fact, just watching a video of tropical fish swimming in an aquarium can lower your blood pressure. Karen Allen, a biopsychologist at the University at Buffalo, found that the blood pressure levels of adults skyrocketed when they had to solve complex mental arithmetic problems in front of their spouses, but barely went up at all when the audience was their pet.

Other studies also provide support for the notion that pets are good for people. For example, children raised in homes with animals are less likely to suffer from asthma and miss fewer days of school because of illness. Elderly people who live with pets have lower levels of cholesterol and depression and elevated psychological well-being. A study of over 10,000 Germans and Australians found that the participants who lived with pets made fewer doctor visits than non–pet owners. Researchers from the University of Melbourne and from Beijing Normal University found that Chinese women who owned dogs slept more soundly, felt better, and took fewer sick days than a comparison group of non–pet owners. A study at the University of Missouri found that a dog-walking program increased the level of physical activity of the participants even outside the program. Some studies show that adults who are attached to animals are less lonely, and there is talk in some medical circles about prescribing dachshunds and Yorkies to old people.

BUT . . . PETS ARE NOT PANACEAS

It would be nice to think that getting a dog or cat would cure all your ills, but don't throw out your Lipitor and Prozac just yet. Media reports on the miraculous healing powers of pets can be misleading. For example, an article in my local paper recently claimed researchers at the University of Missouri found that thirty cancer patients undergoing radiation therapy "considered their health to be improved" after four weekly visits with a hospital therapy dog. The paper got it wrong. First, there were only ten subjects in the pet therapy group; the other participants were in control groups that either quietly read a book during the sessions or talked to a person. More importantly, researchers actually reported that a dozen sessions with a therapy dog was no more beneficial to the cancer patients than reading a book. They concluded there was no association between the dog visits and the patients' mood or perceived health.

This is not the only study in which companion animals had no effect on human health. Deborah Wells, a psychologist at Queens University in Belfast, for example, examined the benefits of pet ownership on a group of people suffering from chronic fatigue syndrome. The pet owners did believe that living with an animal provided them with a wide range of psychological and physical benefits. But objective measures of their health and psychological well-being showed that the pet owners were just as tired, depressed, stressed, worried, and unhappy as the chronic fatigue suffers who did not live with a pet.

The study I mentioned of 10,000 Germans and Australians found that owning a pet had no effect on people's life satisfaction. And an attempt to replicate the boa constrictor blood pressure study failed to reproduce the findings of the original experiment. The effects of dog-walking on fitness are also unclear. The University of Missouri dog-walking study found that while the exercise program was enjoyable for the participants, it did not lower their blood pressure or cause them to lose weight. And a large New Zealand study reported that while new dog owners walked more, their total weekly allotment of physical activity actually went down after they got their pet. In fact, a study of 21,000 people in Finland found that pet owners had higher blood pressure and cholesterol levels than non–pet owners. The pet owners in the Finnish study were also more susceptible to kidney disease, arthritis, sciatica, migraines, depression, and panic attacks. And, while the Finnish pet owners did not smoke and drink as much as non–pet owners, they exercised less often and were more likely to be overweight.

Researchers at the Australian National University found that adults between the ages of sixty and sixty-four who lived with a pet were more depressed, consumed more pain medications, and were in worse mental and physical shape than people without pets. Another study compared elderly people who regularly played with pets with a control group who rarely or never played with animals. The death rates in two groups were the same, and playing with pets made no difference in the participants' health and sense of well-being. (The people who played with pets did, however, drink more alcohol.)

Well, even if owning a pet won't necessarily cure your ills, at least they will make you happier and less lonely, right? Not necessarily. A survey of 3,000 randomly selected American adults conducted by the Pew Research Center in 2006 found that dog owners, cat owners, and people who had no pets at all were equally likely to say they were "very happy" with their lives. Researchers at the University of Warwick in Great Britain examined the effects of acquiring a pet on loneliness in adults. Their subjects completed a psychological test that measured loneliness right after they got their new pet and again six months later. The results were clear: Living with an animal did not make the participants any less lonely.

What are we to make of these disparate findings? Are pets good for people or aren't they? Erika Friedmann recently tried to find out by carefully evaluating the results of thirty studies published between 1990 and 2007 on the effects of living with pets. She found that nineteen of the studies supported the pets-are-good-for-people hypothesis, but that ten of them found that pets had either no effects or had negative effects on human health and well-being. When I discussed these results with Erika, she said, "Yes, pets can be good for people." Then she added, "But they are not panaceas."

WHY ARE PETS GOOD FOR (SOME) PEOPLE?

So pet ownership does seem to make some, but not all, people feel healthier and happier, though perhaps not to the degree the pet industry would have you believe. The scientific question then becomes: Why do pets make a difference? There are three possibilities. The first is that pets don't actually *make* people better off; the causal arrow might point in the other direction. That is, happier and healthier people could be more inclined to keep pets. Maybe they have more money and can afford an animal or perhaps they are in better shape and thus have the energy to take their dog for a walk. A second possibility is that pets do enhance their owners' health and sense of well-being, because pets encourage people to socialize with other humans. For example, Deborah Wells has found that walking a dog facilitates conversations with strangers. The effect, however, depends on the type of dog. The Lab puppies in her study were great social lubricants; the adult rottweilers were not.

The third possibility is that the human-animal bond does actually cause better health by providing social support. To prove this idea, researchers would have to conduct a randomized clinical trial. This involves assigning people who do not have a pet to either an experimental group that gets an animal, or a control group that does not. A randomized clinical trial can be difficult to pull off in the real world, but Karen Allen did it—with rich stockbrokers, no less.

Her subjects were stressed-out Wall Street types who had high blood pressure. At the beginning of the study, all of the subjects were put on an antihypertension drug. Subjects in the experimental group also adopted a dog or cat from an animal shelter while the control group just got the drug. Six months later, Allen and her colleagues put the subjects in stressful situations; one involved a taking a difficult math test, the other giving a speech. The results were impressive. As expected, the blood pressure of all the subjects increased during the stress tests. However, blood pressure went up only half as much in the pet owners as in the drug-only group. Further, the beneficial effects of pets were greatest for the stockbrokers who had the fewest human friends. This experiment is the strongest evidence we have that the presence of an animal in a person's life actually improves their cardiovascular functioning over long periods of time.

PETS CAN BE HAZARDOUS TO YOUR HEALTH

After hearing a stream of his economic advisors tell him, "On the one hand X is true, but on the other hand, Y is true," Harry S. Truman is reported to have muttered in frustration, "Get me a one-armed economist." Anthrozoology has the same problem.

My neighbor Anne recently broke her shoulder when she fell down a flight of stairs after tripping over her dog. This type of pet-induced accident is surprisingly common. Over a year and a half, sixteen seniors were brought to a Sydney emergency room for the treatment of fractures caused by their pets. Among their injuries were four broken pelvises, two broken hips, three broken arms, two broken wrists, one broken ankle, two broken ribs, a broken nose, and a broken neck. The Centers for Disease Control estimates that over 85,000 Americans are injured each year in accidents caused by tripping over their pets, usually dogs.

Pets can be health hazards in other ways as well. A 1999 survey by the Opinion Research Corporation found that nearly one in six dog owners reported having an automobile accident or a close call caused by their pet jumping around in their car. And 60% of the pathogens that humans are susceptible to are zoonotic, meaning they can be contracted from animals. Humans can pick up a cornucopia of nasty diseases from their pets, among them roundworms, skin mites, Lyme disease, brucellosis, ringworm, giardia, leptospirosis, *E. coli*, hookworm, and the aptly named cat scratch fever. When I was twelve, I had a pet baby turtle; everyone did. Who knew that 85% of them carried *Salmonella*? The FDA banned baby turtle sales in 1975, but now snakes, lizards, and other reptiles are popular pets. Predictably, pet-related *Salmonella* is on the rise, and 75,000 cases of *Salmonella* infections each year are contracted from reptiles and amphibians living in American homes. Even therapy animals are a potential health hazard. Several research teams have reported that therapy dogs can acquire and spread MSRA, a serious antibiotic-resistant *Staphylococcus* infection, from patient to patient in hospitals and nursing homes.

IS PET LOVE IN OUR GENES?

Twenty years of anthrozoological research have shown that there are substantial benefits—but also some costs—to living with a pet. These findings, however, do not address the evolutionary mystery of why humans become so attached to animals. Darwinism implies that, directly or indirectly, organisms should act to increase their reproductive fitness—that is, their success in passing down genes to the next generation. But if this is true, why should Joe and his wife pay a thousand dollars a month for chemotherapy to keep their aging golden retriever alive? Wouldn't their genes be better off if they used the money to pay their children's (or grandchildren's) college tuition bills?

On the question of "why pets?" anthrozoologists have offered a wide variety of explanations for the human-animal bond:

- Pets teach kindness and responsibility to children.
- Pets provide "ontological security" in a postmodern age in which traditional values and social networks have broken down.
- Like ornamental gardens, pets are an expression of the human need to dominate nature.
- Pets allow the middle class to pretend they are rich.
- Pets substitute for human friends.

- Pets and people are autonomous beings who gain mutual comfort and enjoyment from their interactions.

All of these could be partially right. But I am most intrigued by a different level of explanation, the evolutionary level. Dan Gilbert of Harvard University claims that every psychologist who puts pen to paper takes a vow to someday write a sentence that begins, "The human being is the only animal that . . ."

Here is mine: "The human being is the only animal that keeps members of other species for extended periods of time purely for enjoyment." The question of why we keep pets is an evolutionary mystery, right up there with why humans are the only mammal with complex symbolic languages, moral codes, religious beliefs, and the ability to learn to enjoy the burn of red hot chili peppers (the spice, not the band).

But as with most "humans are the only . . ." sentences, there are exceptions to mine. There are lots of instances in which a non-human animal has become attached to a member of another species. In the wake of the 2004 Indian Ocean tsunami, an orphaned 600-pound baby hippopotamus named Owen became bonded to Mzee, a 160-year-old giant tortoise in a Kenyan game park. More recently, Tarra, a four-ton Asian elephant at the Elephant Sanctuary in Hohenwald, Tennessee, became fast friends with a rescue dog named Bella. The pair were practically inseparable, and when Bella fell ill, Tarra stood watch for several weeks outside the building where her canine pal was being kept. A comparative psychologist from the University of California at Davis named Bill Mason systematically studied attachments between members of different species by raising young rhesus monkeys with adult dogs. Within a few hours after being introduced, the monkey-dog couples became intensely attached. After a couple of months, each monkey was given a choice between playing with its dog, a strange dog, or another monkey. They chose their dogs. They had become friends.

But attachments between animals of different species almost always occur in unnatural circumstances. There is, for instance, no evidence that, in the wild, our closest relative, the chimpanzee, keeps members of other species around to play with. I remain confident in my statement that humans are the only animal to keep pets.

When and why did this phenomenon come to be? On when, we don't have a clue. Archaeological evidence of pet-keeping goes back 12,000 to 14,000 years for dogs and perhaps 9,000 years for cats. It is possible, however, that some of our Paleolithic ancestors, just like members of some existing tribal societies, captured the occasional parrot or bush pig and brought it home as a pet. The problem is that such early bonds between humans and animals would not leave a trace in the archaeological record. If we were to discover, say, the 25,000-year-old fossil remains of a man cradling a baby monkey, we would not be able to tell if the animal was the dead guy's pet or if the monkey was placed in his grave as a snack in the afterlife.

Lacking any solid evidence of when bonds between humans and animals first formed, the best we can do is guess. Individuals that looked like us were living in Africa 100,000 years ago. However, many anthropologists believe that the real sea change in human thinking occurred roughly 50,000 years ago as evidenced by an explosion of cultural forms: art, music, weaponry, and tools that were exquisite in form and function. Mike Tomasello of the Max Planck Institute for Evolutionary Anthropology argues that this jump in human creativity was fueled by the appearance of a new and sophisticated mental skill—the ability to infer the mental states of other people. Images inscribed on the walls of caves depicting creatures that were half animal/half human suggest that our ancestors began thinking of animals anthropomorphically 35,000 or 40,000 years ago. James Serpell believes that the ability to think of animals as we would a person opened the door for taming wild creatures and forming bonds with them. Serpell makes a good case, but without a time machine, we may never know when the first person decided that a ball of fur could be a friend rather than a meal.

IS PET-KEEPING AN EVOLVED ADAPTATION?

As creationists gleefully point out, we Darwinists argue a lot. We don't fight over whether humans evolved from apes or whether the earth is billions of years old. These are facts. Rather, we squabble over the details. The question of why people love pets is related to one of the most contentious debates in evolutionary theory—the issue of adaptation. Advocates of the adaptationist school of thought are convinced that the human mind evolved to give our Stone Age ancestors a leg up in the Darwinian gotcha game of Who Can Pass Down the Most Genes. They believe that the process of natural selection equipped the human brain with specialized modules for skills like learning language, avoiding sex with close relatives, detecting snakes and cheats, and impressing potential mates.

Critics of the adaptationist paradigm argue that some aspects of human nature could have evolved even though they had absolutely no effect on a person's reproductive success. They think some traits are simply nonfunctional side effects. For example, our bones are white because they are made of calcium, not because women are attracted to men with pallid skeletons. Harvard's Stephen Jay Gould likened nonfunctional biological traits to spandrels, the leftover spaces in a building that architects integrate into the design for aesthetic rather than functional reasons.

Consider as an example of this debate, how adaptationists and non-adaptationists explain the evolution of orgasm in human females. Adaptationists (and I was once among their number) have dreamed up nearly two dozen theories to explain why orgasms occur often in human females but rarely, if at all, in other species. Among their more innovative suggestions are that orgasmic contractions suck sperm up into the uterus, that orgasms evolved as a signal to help women tell which men have good genes and which ones are evolutionary losers, and—my favorite—that by making them woozy, the throes of orgasm keeps women lying down after sex so sperm don't have to swim uphill to reach the egg. Skeptics of adaptationist thinking scoff at these ideas. They explain orgasm in women as the side effect of the fact that orgasm has reproductive benefits for men—just like the presence of nipples on men is the nonfunctional by-product of the fact that nipples evolved so that female mammals can feed their infants.

The argument over evolutionary adaptation also applies to explanations of the human-animal bond. I suspect that most people—including many anthrozoologists—want to believe that love of pets is an attribute of human nature that evolved because it helped our ancestors survive and reproduce. Evolutionary psychologists argue that if a trait is an evolved adaptation, it should be common, widespread, and perhaps, like language, universal. This is not the case with pets. The anthropologist Donald Brown of the University of California at Santa Barbara compiled a list of nearly 400 human universals that ranged from thumb-sucking to beliefs about death. "Interest in bioforms" makes the list, but pet-keeping is conspicuously absent. In many parts of the world most people do not form close bonds with animals. This is particularly true in Africa. My anthropologist friend Nyaga Mwaniki is from rural Kenya. In the village where he was born, people never become attached to individual animals. Indeed, there is no word for pet in Kiambu, his native language. The villagers do keep dogs to guard against intruders and to chase elephants from their gardens. But they never allow dogs in the house, they do not think of them as companions, and they would be horrified at the idea of letting one sleep in their bed.

The argument that pet-keeping is an evolved trait would be strengthened if we had evidence that love for animals has a genetic basis. But we don't. Behavior geneticists use twins to determine the relative influence of genes and environment on a trait. Identical twin pairs will be more alike than pairs of fraternal twins if a trait is strongly affected by genes. By comparing the two types of twins, scientists have discovered that genes are responsible for 90% of the differences between people in height, 50% of differences in how happy they are, and 35% of differences in how often women have orgasms. No one, however, has investigated whether identical twins are more similar than fraternal twins in their attachment to pets. In fact, the role of genes in any aspect of our relationships with animals remains an open question. (The exception is our desire to eat their flesh.)

Finally, if pet-keeping is an evolved adaptation, at some point in human history, people bonded to individual animals must have been better at passing on their genes than their less pet-o-philic peers. Your final score in the game of Darwinism is measured in terms of reproductive success. The fact that your cat makes you happier, healthier, or even helps you live longer is irrelevant. Could living with a pet increase your reproductive fitness? Perhaps girls who grow up with pets are more successful at raising offspring in adulthood because they learn parenting skills by taking care of the family dog. Or maybe early humans who kept pets were more apt to survive the hard times because they could eat their companion animals. I suppose it is remotely possible that some women are turned on by macho guys with big dogs and others by kinder and gentler men who demonstrate their sincerity by cuddling puppies. (Recall that Antoine, the handsome Frenchman, got more dates when he had a dog with him.) But I am skeptical that the possible reproductive advantages our ancestors accrued by falling in love with an animal outweighed the costs in time and resources.

PETS ARE PARASITES?

If pet-keeping does not have an evolutionary function, why do we form such close bonds with animals and invest so much money and emotional energy in them? One possibility is that love of pets is, like the color of our bones, an evolutionary side effect. Consider Harvard evolutionary psychologist Steven Pinker's theory of music. Pinker is an adaptationist when it comes to language and fear of snakes, but he believes that our love of music is a biologically useless consequence of the way that our brains are wired. Could Pinker's view that music has no adaptive value also apply to our love of pets?

Consider the most common claim that Americans make about their pets: "They are my children." Humans are instinctively drawn to animals that remind them of infants—creatures with big eyes, large heads, and soft features. These traits hook our parental instincts and serve to help us pass down our genes by eliciting care for creatures with whom we do share genes: our offspring. But because instincts operate automatically, they can be hijacked. Take, for example, nest parasitism, a reproductive strategy used by dozens of species of birds. A brown-headed cowbird will lay an egg in the nest of an Eastern phoebe. The hapless phoebe will blithely brood the cowbird's egg and then feed the parasitic nestling until it fledges and flies off. Ironically, the phoebe probably gets great emotional satisfaction by raising her faux offspring, not realizing that she has been the victim of a Darwinian sting operation.

In 2005, Mary Jean and I were victims of this scam. The parasite was none other than our cat, Tilly, and the perp was the savvy mother cat who deposited a baby at our door, never to be seen again. Our yellow Lab, Tsali, had died the year before, but we were not looking for a new pet. When I came home from work one afternoon, Mary Jean greeted me with a big smile. That's when I heard a plaintive *meow* coming from the living room. She had found a kitten under our deck. The little sad-sack of a cat had the full complement of baby releasers. With those big eyes and soft fur, she was irresistible. It was a done deal.

The idea that the human-animal bond is caused by a misfiring of our parental instincts appeals to me. The problem is that it does not explain the large cultural differences that exist in the frequency and styles of pet-keeping. Perhaps a different type of evolution—cultural evolution— provides a better perspective on why we love our pets than does Darwin's theory.

PET-KEEPING AS A MENTAL VIRUS

The high point of my intellectual career may have been in 1979, when I found myself sitting next to Richard Dawkins on a bus filled with ethologists who were headed for the Vancouver Aquarium. I had just finished his book *The Selfish Gene* and was star-struck. The book was important for a lot of reasons, but it was chapter 11 that spun my head around. In it, Dawkins argued that evolutionary change does not require either genes or organisms. All you need are replicators—gizmos that can make copies of themselves and have the attributes of longevity, fecundity, and copying fidelity. In biological evolution, the gizmos are the molecular spiral staircases that we call genes that use our bodies to reproduce themselves. Dawkins's insight was that cultural evolution works in the same way. Only with culture, the gizmos are bits of information that are transmitted by imitation and that use our minds to reproduce themselves. The term *gizmo* does not have much scientific gravitas, so he called his hypothetical units of cultural transmission *memes,* a term he coined because it rhymed with genes and harked back to the Greek word for memory.

Memes are everywhere. Some are trivial (wearing baseball hats with their brims pointed backward), some are tragic (a short-lived fad in Japan for strangers to commit suicide together by igniting a charcoal brazier in a sealed minivan), some are transcendent (the arts). Snatches of songs you can't get out of your head are memes. So are cool sneakers and political ideologies. Our ancestors spread memes by copying each other's actions. But memes became a much bigger factor in human evolution with the development of symbolic language. Now, memes spread around the globe at warp speed via radio and television, and, of course, the Internet. The newest mode of mimetic transmission is text-speak, a language I find impenetrable—plz!

The term *meme* is itself an extraordinarily successful meme. When I Googled "meme" ten minutes ago, I got 350 million hits. The word has made it into the *Oxford English Dictionary.* Last night, a TV commentator described a nasty political smear campaign as a meme. But while nerdish techies, edgy hipsters, and a few philosophers have embraced Dawkins's idea, anthropologists, the real experts in cultural evolution, are lukewarm at best. They claim the definition of memes is too loosey-goosey, that— unlike genes—memes are not discrete units. They argue that human cultures march along just fine without the assistance of imaginary replicators.

I am agnostic as to whether memes exist as actual entities. But there is no doubt that ideas and behaviors are contagious, and I find that memes are a useful metaphor for thinking about the role culture plays in our relationships with other species. From the meme's-eye view, pet-keeping is a mental virus spread by imitation. This idea sounds far-fetched, but the evidence for this perverse hypothesis is surprisingly strong.

First, because memes are transmitted by learning, they tend to run in families. Catholic parents usually have Catholic children and circumcised fathers usually have circumcised sons. Similarly, pets run in families; children raised with pets usually grow up to be pet-owning adults. Further, cat kids usually become cat adults, and dog kids, dog adults.

Second, consistent with the meme hypothesis of pet-keeping, societies differ widely in their attitudes toward and treatment of companion animals. Darryn Knobel of the University of Edinburgh studied patterns of dog ownership on the island of Sri Lanka, a culturally diverse society that has one of the highest densities of dogs in the world. In Sri Lanka, your religion determines whether you have a pet dog. In the capital city of Columbo, 89% of Buddhist homes include a dog, as opposed to 4% of Muslim households. The fact that a Sri Lankan Buddhist is twenty times more likely than a Muslim to own a dog suggests that Islam inoculates believers from infection by puppy-love memes, while Buddhism makes people more susceptible.

In addition, like other forms of cultural change, pet memes can spread rapidly. For centuries, the pets of choice in Japanese homes were goldfish and caged birds. After World War II, however, as the Japanese began to emulate aspects of American culture, dogs became more popular, and a quarter of Japanese homes now include a dog. In China, Chairman Mao felt that pet-keeping was a bourgeois affectation and so pets were banned during the Cultural Revolution. When the prohibition on pets was lifted in the 1990s, companion animals rose nearly as rapidly as the number of Kentucky Fried Chicken franchises in Beijing. Presently, 10% of urban Chinese homes include a pet, and the amount of money the Chinese people spend each year on pets went from nearly zero to a billion dollars in a decade.

PET LOVE AND THE MYTH OF SINGLE CAUSATION

Many scientific disputes result from the erroneous belief that if one explanation of a phenomenon is correct, all the others must be wrong. I prefer simple explanations to complex ones, and I would like to think there is one correct answer to the question "Why do humans keep pets?" But, alas, this is not the case.

In explaining animal behavior, ethologists talk about two different types of questions: proximate and ultimate. This distinction also applies to understanding human-animal relationships. Proximate questions focus on the *hows* of behavior—how they work and develop, their underlying neurological and psychological mechanisms. The idea that attachment to pets is affected by levels of the hormone oxytocin in your bloodstream is a proximate-level explanation of pet-keeping. So is the theory that people love pets because they make us feel needed. Ultimate explanations, in contrast, focus on the *whys* of behavior—what their function is, how they evolved, and whether and how they helped our ancestors survive and pass on their genes. The idea that pet-keeping is the result of the misfiring of human maternal instincts is an ultimate-level explanation.

The important point is that both proximate and ultimate explanations of pet-keeping can be correct. I like having Tilly around for lots of reasons. I initially found Tilly adorable because her giant eyes and infantile features triggered parental instincts that helped my forebears pass their genes on to me. This is an ultimate explanation of pet-keeping. But she is fun to play chase-the-laser-pointer-light with, she makes the house feel less empty when Mary Jean is away, and her athleticism sometimes take my breath away (she can shimmy up a dogwood tree in three seconds flat). She has just plain weaseled her way into my heart. These are proximate explanations of Tilly-love.

Different perspectives on why humans keep pets also reflect the biases of different academic disciplines. Clinical psychologists believe that we live with pets because they make us feel loved. Some biologists say pet-keeping is a form of nest parasitism. And some sociologists claim that pets are purely a human social construction; that's why a puppy can be a family member in Kansas, a pariah in Kenya, and lunch in Korea. The bottom line is that our love for pets, the closest of human-animal relationships, is complex and multilayered. Our pets make us feel needed and can provide psychological support when times get tough. But they can also be social constructions and parasites.

Don't feel bad, Tilly. I love you even if you are a socially constructed parasite. I am, after all, infected with the mental virus that tells people to bring cats and dogs into our homes and think of them as children.

Want another little treat, sweetie?

4

Friends, Foes, and Fashion Statements

THE HUMAN-DOG RELATIONSHIP

> I love playing with dogs. As we age, so many people lose the capacity to play, to have fun and enjoy the moment. I am seventy-eight and dogs keep reminding me how to stay in the moment and to enjoy it. Their smiles, wagging tails, and kisses say it all.
>
> —DR. RUBY R. BENJAMIN, PSYCHOTHERAPIST

> If dogs could talk, it would take all the fun out of owning one.
>
> —BOB DYLAN, SONGWRITER

"Don't get so close to the fence. He will bite you in the ass."

I moved away from the fence.

"He" was Maverick, an animal whose genetic heritage runs 98% wolf and 2% dog. The speaker was Nancy Brown, owner of Full Moon Farm, a sixteen-acre sanctuary for wolf-dogs (she never calls them "wolves") near Black Mountain, North Carolina. Some were rescued from abusive

homes, others were brought to Nancy by wildlife officials or animal control officers. One was given to her by a Maryland couple after their bottle-
fed wolf-dog puppy grew up and in one afternoon inflicted $10,000 worth
of damage to their condo.

Full Moon is ten miles down Highway 9, a two-lane blacktop that runs
though a valley that reminds me of what the Blue Ridge Mountains looked
like thirty years ago, before the gated communities started cropping up.
Drive by the Clear Branch Baptist Church, then follow Rock Creek past
a falling-down barn and the volunteer fire department. Take the left fork,
ignore the dead end sign, go a couple more miles, and turn right on a
rough dirt drive you can hardly see from the road. You know you are at
the right place when you see the sign that says, THIS PRIVATE PROPERTY IS
MAINTAINED FOR THE COMFORT AND SECURITY OF OUR ANIMALS. IF YOU
DON'T LIKE THAT THEN PLEASE GO AWAY.

I am a quarter of a mile from the sanctuary when the animals hear my
car and start to howl—eerie, like an old Western movie. But the howls are
intermixed with *arf, arf*—a sound you would expect to hear from golden
retrievers but not from free-range wolves in Montana or northern Italy.
By the time I switch off my car's engine, I am assaulted by a discordant
choir—seventy animals, skittish, each one shouting "Stranger!" in a mé-
lange of dialects that reflect their various stations on the twisted evolutionary path that led canids from the wild to the tame.

Nancy comes out, cup of coffee in hand, and introduces herself. The
animals are still howling full tilt. She can tell them apart by their voices.
Our conversation is peppered with interruptions: *Hear that? That's an argument. That's Aries. Hi, Guinevere. Shut UP, Autumn!* Some of the wolf-
dogs sit up and pay attention when Nancy hollers at them. But most of
them keep pacing. They are nervous around a stranger, tight as the fifth
string on a banjo. They aren't aggressive, just varying degrees of paranoid.

At first blush, they all look like full-blooded wolves to me. Their coats
range from pure white to brown flecked with black and gray, and they
have an intensity that gets your attention. But Nancy shows me some of
the subtle differences between the high- and low-content animals, and
I start to get it. The ones whose genetic heritage leans more toward dog

have wider faces, thicker ears, stockier legs. They bark more. Blue eyes are a sure sign that dog blood runs in an animal's veins. The "98% pures" have the surly James Dean look that wolf-dog groupies love.

Nancy says that the high contents rarely make good companions, that low-content wolf-dogs are easier to live with. If you get a good low-content pup and train it right, you might be able to put it on a leash, take it for a walk, and play with it. Maybe it won't be constantly on the lookout for a gap in the fence or try to kill your neighbor's cat. In other words, your wolf-dog might make a good a pet.

But Nancy cautions me against stereotyping her babies. Even a high-content animal can make a good pet if paired with the right person. She makes her point by entering the enclosure that Maverick (98% wolf) and his pal Mikey share. In a flash, the big animals turn puppy, romping around Nancy, playing with her, cuddling. The chemistry between the three of them is magical. But even with these, her favorites, Nancy has rules. She never lets them get in a position above her head, and she never plays tug-of-war with them. I ask Nancy how many of her animals have the potential to be rehabbed for family living. She looked up and thought for minute, mentally ticking them off, and said, "Four." Not good odds out of six dozen.

The legal status of wolf-dogs is murky. In North Carolina, anyone can own a wolf-dog, but in some states, they are banned outright. They are classified as wildlife in Pennsylvania, and you need a special permit to keep one on your property. Sandra Piovesan of Salem, Pennsylvania, thought of her nine wolf-dogs as pets rather than exotic wild animals, so she registered them as dogs and treated them like her children. A few weeks after she told one of her neighbors that her wolf-dogs gave her "unqualified love," Sandra's body was found in their enclosure, mauled by her pets.

This was not an isolated incident. Nineteen people were killed by wolf-dogs between 1982 and 2008 in North America, compared to nine by German shepherds. There are, however, a lot more German shepherds than there are wolf-dogs. Merritt Clifton, editor of *Animal People*, an independent newspaper that covers animal issues, says that a wolf-dog is sixty times more likely than a German shepherd to maim or kill a child.

Wolf-dog fans don't buy it. To them, wolf-dogs are misunderstood, their reputation undeserved.

Driving home from Full Moon, I was keyed up. The animals were magnificent and I admired Nancy's commitment to them. If not for her, these animals would have been euthanized. Instead, their lives are as good as it gets for creatures whose heritage is an amalgam of predator and pet. But their edginess rubbed off on me. I felt like a journalist who had spent the day riding with an outlaw motorcycle gang.

Unsettled, I went for a walk that evening and ran into my friend Jeanette and her dog, Bindi, a mellow little mutt she had rescued from an animal shelter. Bindi looked up at me expectantly, and I reached down and scratched her behind the ears. Suddenly I felt relaxed. It was hard to believe that the difference between Bindi and the skittish gray ghosts I had spent the day with boiled down to a few base pairs of DNA.

THE RARIFIED WORLD OF SHOW DOGS

The best way to observe the seemingly infinite variety of forms that humans have sculpted from wolf genes is to spend a couple of days hanging out at an all-breed conformation show sponsored by the American Kennel Club. The Asheville Kennel Club hosts one of these events every summer, and it attracts 1,500 dogs, owners, and professional handlers from around the country. Some fly in, but even more show up in Winnebagos and motor homes. The Western North Carolina Agricultural Center's parking area quickly fills with lawn chairs, gas grills, and portable pens. Vendors hawk dog paraphernalia—the bite-sized pieces of meat handlers use to make dogs perk up in the show ring; mastodon-sized bones for Great Danes and Saint Bernards; assorted grooming gear, lotions, and shampoos; dietary supplements; bows and hair clips for the little dogs; socks embroidered with breed silhouettes for their owners.

In the main arena, judges, stewards, owners, and casual observers mill around, chatting between events, the owners combing, snipping, and flicking a stray hair on a poodle, wiping slobber off a Newfie's face. The dog people look perfectly normal—there are roughly equal numbers of men and women, and they range from retirees to kids getting ready for the Junior Showmanship competition. The biggest surprise is how quiet the arena is. There are over a thousand dogs, each waiting for his or her turn in the spotlight, but I don't see any poop and, with the exception of a couple of yappy Chihuahuas, there is little barking. These dogs are professionals.

I wander around behind the scenes, taking photographs and asking people about their dogs. Like most of the animal people I have met over the years, their eyes light up when they talk about their dogs. It's like asking parents about their children. Dog people are easy to talk to, and their enthusiasm for their dogs is infectious.

There is, however, the occasional weirdness at these events. I notice a well-dressed woman, a professional handler, sitting next to the tawny Great Dane she will soon be parading around Ring Four. On the table in front of her is a pile of gummy bears—the chewy rainbow candy that kids like. The woman puts a fistful of gummy bears in her mouth and starts to chew. After the candy has transformed into a slimy mess, she reaches in her mouth, extracts the amorphous glob of sugar and spit, and nonchalantly shoves it into the big dog's open maw. Huh?

That afternoon, I mention the gummy bear incident to another handler, a tall blonde woman from New Orleans who is getting ready to show a lovely white and black Japanese chin. "Oh, sure," she tells me. "Professional handlers do that all the time. It helps the dogs get to know you. I use chicken." She points to a four-inch hunk of boiled chicken meat tucked into her armband. As she leads the dog, Fred, into the ring, she tucks the whole thing in her cheek like a wad of Red Man. As the judge approaches her dog, she takes the piece of chicken from her mouth, waves it an inch in front of Fred's nose, then puts it back into her mouth. Fred perks up.

On this day, I am in luck. I run into a woman named Barb Beisel who is grooming a little Havanese. A highly regarded professional dog handler and breeder, she agrees to take me under her wing for the next two days. Barb introduces me to judges and a couple of top handlers: Jimmy Moses, who has been showing star-quality German shepherds for decades and Chris Manelopoulos, whose world revolves around a white standard poodle named Remy. (Seven months later, I watch both of them on national television running their dogs around the ring at Westminster. Remy took the nonsporting group, but, in an upset that made national headlines, he lost to the energetic Uno, the first beagle ever to win Best in Show.)

Even at the relatively small Asheville dog show, most of the handlers are professionals. It will cost you between $50 and $100 each time a handler takes your dog around the ring. Unless you know what you're doing, hiring a professional is money well spent. Barb tells me even a really good dog does not have much chance winning without a pro parading it before the judges.

Barb worked in insurance until she took to the dog circuit full-time. She has silver hair and twinkling eyes and looks a bit like the fairy godmother in *Cinderella*. But when she leads a dog into the ring, she is all business. A lot rides on each animal's performance. High-rollers in the elite world of international dogdom hire Barb to "finish off" their prize pooches. When Barb takes on a new client, a dog she thinks has star power, she first teaches it how to show its stuff in the ring—to look alert, to follow her hand as she waves a morsel of meat in front of its nose, to patiently put up with judges who poke their withers and peer at their gums. Then she carts the dogs around the country until they earn the points needed to be awarded the title Champion. Barb is always on the move, traveling to dog competitions in her motor home along with an entourage that includes a dozen show-quality terriers, her assistant Marie, and an aging ferret.

Barb loves dogs, but not all breeds equally. She is partial to the toys—cute, energetic little animals with high-pitched yips and bows in their hair. While Barb was showing me the fine points of grooming a six-pound Yorkie, another handler walked by leading a mastiff the size of a pony, its testicles nearly as big as the Yorkie's head. Barb glanced at the slobbering brute and muttered, "I just don't know how anyone can love a creature like that." I suspect the mastiff's owner may have been thinking the same thing about the lap dog whose lustrous coat Barb was brushing for the twentieth time that afternoon.

THE PATH FROM WOLF TO WHIPPET

How did wild creatures that looked exactly like Nancy Brown's gang of lupine bandits become transformed into animals as different as a giant mastiff and an elfin Yorkie? How did the descendents of gray wolves become the most variable mammal on earth? A mastiff can reach 200 pounds while an adult tea cup Yorkie can tip the scales at 2 pounds. This difference in size is proportionally larger than that between me and a full-grown African elephant. Molecular biologists Heidi Parker and Elaine Ostrander of the National Institute of Health call the domestication of dogs the most complex and extensive genetic experiment in human history. Remarkably, this morphological miracle took place in the blink of an evolutionary eye.

Charles Darwin thought the modern dog was a mix of coyote, wolf, and jackal. He was wrong. If we know anything for sure about canine evolution, it is that the ancestor of the dog living in your house—whether Pekinese or rottweiler—was a gray wolf. It makes sense that the wolf was the first domesticated animal. Like us, they are social, active during daylight, and good at figuring out who is the boss. Hooking up with odd-looking and hairless two-legged creatures was a strategy that worked out well for *Canus lupus familiaris*; an estimated 400 million dogs are running around the planet right now, compared to a couple hundred thousand gray wolves. But when, where, and why did our ancestors first invite a predator into their lives?

First, when. The answer is not very long ago. Every dog owner I know has a slew of pictures of their pets around the house. If our Stone Age ancestors cared as much for dogs as we do, they probably would have made pictures of them too. But they didn't. Paleolithic cave art is rife with stunning images of reindeer, buffalo, horses, mammoths, ibex, rhinoceroses, bear, lions, deer, and even a few fish and birds. But creatures that look like dogs do not appear in the early Stone Age bestiary. To learn how dogs and humans came to throw in their lots together, we must turn to bones and genes.

Members of the genus *Homo* have lived alongside wolves for 500,000 years, but there are no signs that wolves and early humans were on friendly terms. Evidence of relationships between man and wolfish-looking dogs shows up only recently in the fossil record. Domestication changes a species. First, they get smaller. Juliet Clutton-Brock of the Natural History Museum in London believes this is an adaptation to malnutrition during pregnancy, the result of selection for larger litters of smaller babies. The earliest dogs had smaller jaws with more crowded teeth, wider faces, and shorter snouts than wolves do. Like most domesticated animals, they also had smaller brains than their progenitors. In short, the original dogs looked like puppified wolves.

The fossils suggest that the domestic dog appeared sometime between 14,000 and 17,000 years ago. The first convincing evidence of a human-canine bond is a 12,000-year-old skeleton of an elderly woman unearthed in northern Israel who appears to have been carefully buried cradling a puppy. Dogs quickly became fixtures of late Stone Age cultures in Europe and Asia. By 10,000 years ago, dogs were romping around the New World, having accompanied humans who made their way across the land bridge from Siberia to North America. It only took 4,000 years for dogs to make the long journey from Alaska down to Patagonia.

The tools of the archaeologist's trade are spades, dental picks, and dusty hiking boots, but molecular biologists trying to ferret out the mysteries of canine evolution wear lab coats and spend their days listening to the hum of DNA sequencers. The raw materials for their research are not fossilized bits of jaw and teeth, but fragile strands of genetic material, mitochondrial DNA that has been passed from mother to child over thousands of generations. Mitochondrial DNA is different from the nuclear DNA that makes up the genetic instructions you bestow on your kids when sperm meets egg. Mitochondrial DNA floats in a cell's cytoplasmic goo, converting organic material into energy. Most important for evolutionary biologists, all your mitochondrial DNA comes from your mother. Unlike regular DNA, which is randomly shuffled during sex, ancestral lines of mitochondrial DNA remain unchanged from generation to generation, except for the occasional mutation.

Geneticists use the rate of mutational change in mitochondrial DNA as a kind of molecular clock that gives an indication of when, where, and from what a species has emerged. In 1997, an article appeared in the journal *Science* that shook up the world of people who care about things such as when dogs made the great leap from wolves. A group of researchers led by Robert Wayne of UCLA analyzed the mitochondrial DNA of sixty-seven dog breeds as well as the DNA of wolves, coyotes, and jackals. Based on their interpretation of the molecular clock, they concluded that dogs emerged from gray wolves between 100,000 and 135,000 years ago, ten times earlier than the fossil chronology of the dog-wolf split. That is a huge discrepancy. What's going on? The accuracy of a molecular clock, just like an alarm clock, depends on when you set it, and the UCLA researchers may have been overly generous when calibrating theirs. Most biologists now believe that dogs began to diverge from wolves between 15,000 to 40,000 years ago, a date that jibes better with the story the bones tell.

WERE THE FIRST DOGS PETS OR DUMPSTER DIVERS?

Anthrozoologists disagree about how humans and dogs came to throw in their lots together. One popular theory goes like this: Fred Flintstone, Stone Age hunter, stumbled upon a wolf den one afternoon while foraging for dinner and brought a baby wolf back to the cave to cook for the evening meal. But just as Fred was about to toss little Lobo on the campfire, his wife, Wilma, looked deeply into those adorable puppy eyes and her maternal instincts kicked in. She snatched the puppy from Fred and doted on it as if it were her child, maybe even suckling the pup. The wolf's fierce nature was soon tamed, in Bill Monroe's words, "by a good woman's love." *Voilà*, Lobo became the first domesticated companion animal, and soon produced puppies that, over time, came to look and act like Rin Tin Tin.

Ray Coppinger, a biologist and dog-sled racer who teaches at Hampshire College, is skeptical of this idea, which he dismisses as the "Pinocchio Hypothesis." Coppinger thinks that converting full-blown wolves into dogs is about as likely as the fairy godmother in Pinocchio waving her magic wand over the wooden puppet and turning it into a flesh-and-blood boy. The trouble with the wolf-into-puppy-dog theory, says Coppinger, is that pure-blooded wolves are essentially untamable. That's why you don't see them in circus animal acts. Coppinger admits that they can learn a few tricks and that some become habituated to walking on a leash, but he feels this is a far cry from being truly tame. Other researchers have found that even hand-reared wolves do not become attached to their caretakers the way dogs do. They are more likely to growl and literally bite the hand that feeds them.

But if humans did not tame wolves, how did they become our pals? Coppinger says they tamed themselves. He traces the emergence of dogs to the time people began to give up the nomadic life and settle into permanent villages. Settlements generate piles of refuse, potential gold mines for opportunistic scavengers. Wolves who were less nervous around humans would be more efficient garbage-pickers, just like the dogs-gone-wild you can see today hanging around garbage dumps on the outskirts of Lagos or Mexico City or Istanbul. These animals would have been better nourished than their warier competitors, hence bearing more offspring—puppies that would share their parents' genes for being less fearful of humans. Over generations, these tamer animals would have access to more food and eventually adapt to living around people. This self-domestication process would set the stage for unconscious selective breeding for functional traits like barking at strangers and chasing game.

HUNTING LIZARDS WITH DOGS

Coppinger may be right. I once got a glimpse of how humans may have used the first dogs to hunt small game. My graduate school friend Bev Dugan spent several years in the jungles of Panama figuring out the sexual strategies of green iguanas. She had a problem: All iguanas look more or less like reptilian versions of Keith Richards, and they spend most of their days in treetops. If her project was going to succeed, she needed to capture and mark the animals in order to tell them apart. But how do you catch a creature that lives in the canopy of a tropical forest?

That's where dogs come in.

Another researcher introduced Bev to Pifo and Cesar, two local men who had perfected the art of hunting big lizards. When I visited her research site, they invited me to tag along on one of their forays. Pifo's dog was a brownish mongrel, medium-sized, with short hair, a curly tail, and perky ears—the classic profile of dogs that go feral and are allowed to breed freely. Ray Coppinger calls these "natural breeds." They usually resemble Australian dingos, and you find them running wild from Tanzania and Israel to the coast of South Carolina. Pifo's dog was a companion as well as a hunting partner. He talked to his dog, petted him, and fed him table scraps.

An iguana hunt begins when you spot a lizard on a limb basking in the sun. The acrobatic Cesar shimmies up the tree trunk and gingerly makes his way out on the branch. The dog, tense, stares up at the lizard. Cesar violently shakes the branch until the iguana bails and plunges straight down sixty feet into the underbrush. In a flash, the dog is after it. There is no way a human could chase the lizard down, but thanks to Pifo's dog, *no problema.*

Cesar and Bev take off after the dog. Clumsily, I do my best to keep up. The dog quickly corners the lizard. The trick for Bev is to subdue the iguana before the dog does any damage to it or it does any damage to her. Male iguanas can be six feet long. They will whip you with their tails and bloody your arms with their claws. You do not want to get bitten by one. Bev usually wins these woman-versus-lizard wrestling matches. Once she has the iguana under control, she weighs it, checks its reproductive status, marks it, and lets it go.

It is an efficient way to hunt. With the aid of the dog, they can capture ten or twenty lizards a day. When Pifo and Cesar were not catching iguanas for Bev's research project, they used their dog to hunt lizards for meat. Once they catch an iguana, they immobilize it by yanking a tendon from the iguana's forefeet and use it to tie its legs together behind its back. Later, they will kill the lizard by inserting a long spiky thorn straight through the animal's brain and into the spinal cord.

Coppinger argues that pure-blooded wolves are too wild for this sort of cooperation between man and beast. After spending time with Nancy Brown's wolf-dogs, I am inclined to agree. But the most convincing support for Coppinger's theory of dog domestication comes from an extraordinary genetic experiment with foxes.

FROM WILD TO TAME: HOW TO MAKE FOXES FRIENDLY

In the mid 1950s, a geneticist named Dmitri Belyaev found himself on the losing end of a scientific dispute and was fired from his position at Moscow's Central Research Laboratory of Fur Breeding. Belyaev wound up as head of an animal research facility in Siberia where, in 1959, he began a remarkable experiment on silver foxes that went on for forty years and involved over 45,000 animals. It changed the way scientists think about domestication.

By selectively breeding foxes for tameness around humans, Belyaev sought to create a strain of captive foxes that would be easier to work with. At the fur farm, he tested fox kits for their reactions to people and then cross-bred the tamest males and females. Initially, only a few foxes met his strict friendliness-to-humans criteria. By the tenth generation, 20% of foxes were tame, and after forty generations, 80% of fox kits were tame. The selectively bred foxes would lick your face; the unselected line of animals would rip it off.

So foxes selected for tameness became nice. Big deal. Here is the important part: An unintended by-product of selection for tameness was that the foxes began to look and act like dogs. Over the generations, their ears became floppy and their tails curly. Their coats began to show traces of brown (the probable color of early dogs), and some of them developed the white patches you sometimes see on the faces of dogs. The foxes' faces became shorter, wider, cuter. The animals showed decidedly un-foxlike behaviors. They wagged their tails when people greeted them. The foxes' physiology changed, too. Compared to normal foxes, their levels of the stress hormone cortisol were lower and their neurons produced more of the brain's natural antidepressants. No wonder they were mellow.

Why should traits as different as coat color, the curl of the tail, and the shape of the head be dragged along with having a nice personality? We don't know, but behavior and color are genetically linked in species ranging from garter snakes to rats. Whether the changes that occurred in Belyaev's tame foxes were produced by a small number of linked genes, or even a single gene, is unclear.

CAN DOGS READ OUR MINDS?

When I hear Doc Watson, the great flat picker from Deep Gap, North Carolina, lay out just the right notes in a fiddle tune guitar break, a tingle runs from my left shoulder and up the side of my neck and I get goose bumps. The Hawaiians call this "chicken-skin music." As a scientist, I read a lot of articles in academic journals. Most of them put me to sleep, but every now and then, one gives me the chicken skin. It happened in 2002 when one of my colleagues handed me an unpublished manuscript that his old college roommate, Mike Tomasello, a developmental psychologist, sent him. The paper, which soon appeared in the journal *Science*, compared the ability of wolves, chimps, and dogs to understand human gestures. The lead researcher was a twenty-six-year-old graduate student named Brian Hare.

To understand the importance of the study, you have to know something about chimpanzees. They are our closest, smartest, and most socially savvy relative. Like other nonhuman primates, however, chimps are lousy when it comes to following human signals to find food. Even chimpanzees that are hand-raised by people are not able to use gazes and finger points to locate food. Given that big-brained chimps are poor at this task, you would predict that wolves, which have much smaller brains than apes, would also fail the point-gaze test. You would be right. You would also, of course, think that domestic dogs would be dumber than wolves as their brains are roughly 25% smaller. Wrong.

Hare, who now heads the Canine Cognition Center at Duke University, used a simple choice experiment to show that dogs are naturals at understanding human signals. He hid food under one of two bowls. A research assistant would tap, point, and gaze at the bowl with the hidden food. As expected, the chimps and wolves were hopeless. They randomly guessed the left or right bowl and were right about 50% of the time. The dogs did much better. When given all three cues, the dogs made the right choice 85% of the time.

Is the superior performance of dogs the result of nature or nurture? He believes that evolution has endowed dogs with a special ability to read human signals. Recently, however, the origins of the differences between dogs and wolves have become a matter of hot debate. Clive Wynne, a dog researcher at the University of Florida, thinks Hare is wrong. He attributes the differences to socialization. The truth is probably somewhere in between. A research group in Hungary headed by Adam Miklósi has found that you can train wolf pups to attend to human signals, but it takes the wolves ten times longer to achieve a similar level of success and requires much more socialization.

An increasing number of scientists are focusing their research on the canine mind, and dog behavior research centers are popping up everywhere. The remarkable ability of pet dogs to understand human intentions and their eagerness to participate in experiments makes them excellent subjects for studies of the evolution and development of communication and social behavior. As Yale University's Paul Bloom says, in cognitive ethology, dogs have become the new chimpanzee.

Unexpected examples of canine brain power appear every month in journals like *Science, Animal Behavior, The Journal of Comparative Psychology*, and *Animal Cognition*. A border collie named Rico tested by researchers in Germany knew the meaning of over 300 words and could learn the name of a new object in a single trial. Scientists in Brazil taught a dog to use a keyboard so she could indicate whether she wanted to play or eat, be petted, or go for a walk. Researchers at the University of Vienna found that dogs can tell when their owners are watching them or watching television— and they disobey more when their owners are glued to the TV set. Further, dogs are copycats. For instance, when their owners yawn, they tend to yawn. And when they try to solve a problem, dogs imitate the behavior of humans whose actions succeed rather than that of those whose actions fail.

WHAT WOULD LASSIE DO?

Studies like these show that dogs have some amazing abilities. But do we sometimes expect too much from an animal whose brain weighs a couple of ounces? For thousands of years, humans have used the intelligence of dogs to good effect—watch a border collie bring in a flock of sheep or a gun dog on point. In addition to hunting and herding, we now count on dogs to sniff out bombs, dope, and bladder cancer. They locate lost kids and rotting cadavers, warn deaf owners when the smoke alarm goes off, and lead the blind through city streets, even "intelligently disobeying" commands that would jeopardize their owners. A lot of people go further and believe that their dogs possess quasi-mystical abilities. Half of dog owners believe that their pets have a form of ESP that allows them to know in advance when they are about to return home after being away.

Humans love the idea that a good dog knows when you are in a jam and will help you get out of it. When I was a kid, Lassie was my idea of what a pet should be. Bosco, our beagle, was fun to play with, but Lassie could save you from a cougar, fetch your dad if you fell down a well, and bark when the barn caught fire. (Many people still feel this way. Just Google "dog saves owner.")

Bill Roberts, a psychologist at the University of Western Ontario, was impressed by news accounts of dogs who rescue people. Bill knew that social psychologists investigate the circumstances in which human by-standers come to the aid of strangers by staging fake accidents. He thought that similar experiments might shed light on a dog's ability to intentionally seek help when their owners are in dire circumstances. Fortuitously, one of his students, Krista Macpherson, was a dog breeder and trainer. She had access to lots of dogs and owners, and she and Bill came up with a way to test what I call the Lassie Get Help hypothesis.

They asked dog owners to walk their pets through a field in which a bystander sat quietly in a chair reading a book. (The bystander was a part of the research team.) When the dog-owner pairs were thirty feet from the bystander, the owner would clutch his chest, collapse, and lie motionless on the ground. When faced with this situation, Lassie, of course, would sense that her owner was in dire straits, bark once or twice, and run over to the bystander and tug at his sleeve. In this situation, did the real dogs—dogs like my beagle Bosco—step up to the plate?

No. Not one of the animals in the experiment made the slightest effort to solicit help for their injured owner. Nor did they try to get help when Bill repeated the experiment in a laboratory situation in which a fake bookshelf was rigged to fall and pin their owners to the floor. Even the collies in Bill's study failed the Lassie Get Help test.

Some breeds, however, are more adept than others at understanding humans and responding to human commands. For example, Adam Miklósi's group in Budapest has found that breeds that have been selectively bred to work with humans, such as sheepdogs and retrievers, are better at using human hand points to locate food than breeds like hounds that were bred to work independently. Not too surprising. What is surprising is that breeds with short snouts and wide faces like bulldogs, boxers, and pugs understand human signals better than long-nosed breeds like Dobermans, dachshunds, and greyhounds. The researchers suspect that this odd finding boils down to a difference in the placement of the eyes and the shape of the head. The eyes of the narrow-faced breeds are located more on the sides of their heads. While this trait gives them better peripheral vision, it also makes them more distractible.

Researchers at the Center for the Interaction of Animals and Society at the University of Pennsylvania have examined breed differences in dog behaviors in a different way. They developed a Web-based owner survey to examine traits like the willingness to obey commands. Their Canine Behavioral Assessment and Research Questionnaire (C-BARQ for short) has been taken by thousands of dog owners and has provided a goldmine of data on differences in the behaviors of dog breeds. (If you have a dog and would like to participate in C-BARQ research, the survey Web site is w3.vet.upenn.edu/cbarq.)

The researchers were particularly interested in a dog's willingness to do what its owner wanted it to do—come when called, fetch, learn tricks, and pay attention. The researchers analyzed C-BARQ trainability scores for 1,500 dogs representing eleven breeds. They found that some types of dogs are easy to train while others are not politically smart. Labrador retrievers were the most trainable breed and basset hounds the least. Seventy percent of Labs were highly trainable, compared to only about 5% of basset hounds. And while none of the Labs were in the dense categories, half the basset hounds were.

Big differences also show up in the frequency with which different breeds of dogs cause harm to humans. Because they are "man's best friend," it is easy to forget that dogs are one step removed from an animal that kills for a living. Four and a half million Americans are bitten by dogs each year, and over 800,000 of these bites require medical attention. Many are children who have been bitten in the face. About 70% of serious attacks are attributed to just two breeds: pit bulls and rottweilers. On the other hand, Chihuahuas and dachshunds are responsible for just one half of one percent of human dog-bite injuries. Pit bulls and rotties carry the bad seed while Chihuahuas and dachshunds are innately nice, right?

The University of Pennsylvania researchers used the C-BARQ to evaluate the propensity of 4,000 dogs from thirty-three breeds to bite strangers, turn on their owners, and pick fights with other dogs. Surprisingly, pit bulls and rottweilers were not the breeds most likely to attack people— Chihuahuas and dachshunds were. Pit bulls and rotties ranked in the middle of the pack. They were about as aggressive as poodles.

BAN THE BREED OR THE DEED?

Differences between breeds in serious attacks on humans raise one of the most contentious issues in our relationships with dogs: Are some breeds so inherently dangerous that it should be illegal to own one? Pit bulls were once considered a model of courage and loyalty, an excellent family pet. A pit bull named Petey was a central character in the *Little Rascals* films. During World War I, pit bulls were featured on military recruiting posters. But in January 2009, the United States Army, in response to six fatal dog attacks in base housing, issued a directive prohibiting pit bulls and other "aggressive or potentially aggressive breeds" on military bases.

Denver has banned pit bulls outright. In Ohio, all pit bulls are legally considered "vicious animals," even ones that have never shown any sign of aggression. State law in Ohio mandates that pit bulls be confined in a secure locked pen, restrained on a leash, and that their owners maintain a $100,000 liability insurance policy. Because most pit bull owners ignored the law, the city of Cincinnati has outlawed the breed entirely.

Other communities, however, in response to pressure from animal welfare advocates and pit bull enthusiasts, are rescinding laws that discriminate against specific breeds. The mantra of the pro–pit bull faction says, "Ban the deed, not the breed!" This battle came to a head in Seattle when a group of residents who had run-ins with pit bulls approached the city council about enacting legislation that would restrict the ownership of dangerous breeds. On September 8, 2008, the *Seattle Post-Intelligencer* ran a story on the efforts of local pit bull defenders to stave off efforts to outlaw their pets. The timing was unfortunate. While *Post-Intelligencer* readers were sipping their morning coffee and reading about what loving pets pit bulls are, a seventy-one-year old Vietnamese woman named Houng Le who lived in Sea Tac, a Seattle suburb, was being worked over by a pair of them. Her neighbor grabbed a pitchfork and tried to fend the dogs off of Mrs. Le, but the attacks did not stop until a couple of deputy sheriffs showed up and shot the dogs. After ten hours, surgeons were able to successfully reattach Houng Le's ears, but were unable to repair her mangled right arm. According to neighbors, the animals were well-behaved and gentle. One of them said, "They were playful and would lick you silly."

Bans against "vicious breeds," particularly pit bulls, have become one of the most heated and divisive animal-related issues in the United States. Pit bull supporters are completely devoted to their breed. They argue that the breed's outlaw reputation is not deserved, that most pit bulls never bite anyone, and that breed-specific legislation is a canine version of racial profiling. They want to rebrand pit bulls as "America's Dog." Pit bull detractors are equally vehement. They tell you that between 1982 and 2008, pit bulls were responsible for 700 maimings and 129 human deaths in the United States and Canada. (In contrast, during the same period Labrador retrievers, the most popular dog in America, maimed twenty four people and killed three.) Anthrozoologists are divided on breed bans. Most researchers I know oppose laws that would keep people from choosing a dog because the breed is inherently dangerous. Not everyone agrees. For example, Alan Beck, a pioneer in anthrozoology and director of the Center for the Human-Animal Bond at Purdue University, supports breed bans. He thinks pit bulls are loaded guns. And Sarah Knight, a British anthrozoologist, wants to see the breed become extinct.

The same split exists among animal protection organizations. The American Veterinary Medical Association, the American Society for the Prevention of Cruelty to Animals, the Humane Society of the United States, and the American Humane Association lobby against breed-specific legislation. You might think that the more radical PETA would also put their moral muscle and political clout on the side of the dogs. Not true. PETA wants pit bulls gone. They argue that pit bull bans are about reducing suffering, not racial profiling. PETA points out that pit bulls are the victims of much of the abuse that Americans heap upon dogs. Organized dogfights are only part of the problem. PETA claims that pit bulls are more likely than any other breed to be beaten, starved, and neglected by irresponsible owners. Several studies have linked pit bull ownership with antisocial behavior. This includes an Ohio study which found that owners of high-risk dogs, almost all of whom were pit bulls, were seven times more likely to have been convicted of violent crimes and eight times more likely to have been convicted of drug offenses than owners of low-risk dogs. Finally, 900,000 unwanted and unadoptable pit bulls are put to death in animal shelters in the United States each year.

Keep in mind, however, that the C-BARQ study found that pit bulls and rottweilers are no more vicious than the average beagle. Only 7% of pit bull owners reported that their pet had tried to bite a stranger, compared to 20% of dachshund owners. So why are these breeds responsible for two-thirds of fatal dog attacks? I have to admit that I would rather have it out with an angry Chihuahua than a rottweiler or pit bull who wanted to eat my lunch. Even if they are less aggressive than Chihuahuas, an attack by a rottweiler or pit bull is much more likely to put you in the hospital simply because the dogs are big and powerful and can do more damage.

One reason for the jump in rottweiler and pit bull attacks on humans is that the breeds suddenly became popular. In 1979, most people had never heard of a rottweiler. The breed ranked forty-first in popularity in the United States, with only 3,000 new puppies registered with the AKC each year. In contrast, with 60,000 puppy registrations a year, German shepherds were among the most loved dogs in America. But in the mid-1980s, while German shepherd registrations remained essentially flat, rottweilers became hot. Between 1979 and 1993, rottie registrations soared to over 100,000 new puppies a year. They became the second most popular breed in the United States, and nearly a million registered rotties were living in American homes.

The surge in the demand for rottweilers came at a cost. Between 1979 and 1990, rottweilers killed six people in the United States compared to thirteen fatalities by German shepherds. Abruptly, however, the gruesome death statistics reversed. Over the next eight years, German shepherds killed four people while rottweilers killed thirty-three. The spike in deaths was, in part, attributable to their increased numbers. Even if a breed's temperament is unchanged, more dogs mean more dog attacks.

In 1993, rottweilers temporarily overtook pit bulls as America's most dangerous dog. The breed, however, soon became the victim of its own popularity. The rash of attacks generated negative publicity, and insurance companies began to cancel the policies of rottweiler owners. Over the next decade, rottweiler registrations tanked, falling from 100,000 puppies a year to fewer than 20,000. With the right owners, rotties, like pit bulls, can make great pets. But a lot of people who impulsively picked up a rottweiler puppy at the height of their popularity were unprepared when their cute bundle of fur morphed into Cujo.

OODLES OF POODLES: WHY DOG BREEDS
SUDDENLY BECOME POPULAR

The rapid rise and fall in the popularity of some dog breeds raises the more general question of what fuels changes in human cultures. Take Crocs, those ugly, squishy plastic shoes. Did Crocs become popular because they were in some objective sense superior to other shoes—cheaper, more comfortable, better for your back? Or were they simply a fad that swept across American culture like the flu?

We can ask the same question about the sudden popularity of breeds like rottweilers. After all, rottweilers are expensive to feed, have to be delivered by C-section, and are prone to hip dysplasia, diabetes, cataracts, and Addison's disease. Theirs was not, however, the biggest fad in American canine history. That distinction belongs to poodles. Between 1946 and 2007, 5.5 million poodle puppies were registered with the AKC, 2 million more than the runner-up, Labrador retrievers. Poodle popularity peaked in 1969, when nearly a third of all new registrations were for poodles and the AKC hired a full-time staffer just to handle the glut of poodle paperwork. The ascendancy of the breed was fast; yearly registrations increased 12,000% between 1949 and 1969. The poodle craze was not limited to dogs. Poodle skirts, usually white or pink and embellished with the silhouette of a French poodle, became *de rigeur* for the bobby-sox crowd. Today, they are hot items on eBay.

Consider now the English toy spaniel. While poodles were invading postwar American culture like army ants on the march, toy spaniel registrations dropped from a measly 123 in 1949 to an even lower 45 in 1969. These dogs are small and cute. They look like Lady in *Lady and the Tramp*. According to the official AKC breed description, they are "a bright and interested little dog, affectionate and willing to please." What could be better than that?

The difference in the popularity of these two breeds raises an issue that transcends our taste in pets. Why do some foods, songs, colors, books, religions, sneaker styles—you name it—catch on while others languish in obscurity?

One possibility is that, like a genetic mutation that increases an organism's reproductive fitness, some cultural innovations become spectacularly successful because they are better than the competition at doing something. Pop-top beer cans and iPods come to mind. But transient enthusiasms for nose rings and fondue pots are more like nonfunctional variations on a theme. Did poodles come to dominate the dog niche in American homes because they were smarter or more obedient or easier to fall in love with than the unpopular breeds? Or were they just famous for being famous?

I became interested in this question when I was asked to write an article on the evolutionary psychology of human-animal relationships and decided to include a discussion of cultural evolution. I was trolling the Internet for examples of rapid changes in attitudes toward animals when I stumbled on the American Kennel Club Web site, which listed the numbers of puppy registrations for each breed for the previous three years. Scanning the columns of figures, I noticed that the number of Dalmatian puppies had dropped precipitously. I called a Dalmatian breeder I knew who was a member of the AKC board of directors and asked if she could arrange for the organization to give me Dalmatian registration numbers over a longer period of time. A few weeks later, a thick package from the AKC's Manhattan headquarters showed up at my office. It was the mother lode—sixty years of registration statistics for every AKC breed. Forty-eight million dogs.

That's the good news. The bad news was that I now had to make sense of the largest data set in the history of psychology. Not knowing what else to do, I made graphs charting the growth of each breed. I found that since the end of World War II, only four breeds have been America's most popular dog. They are, in order: cocker spaniels, beagles, poodles, cocker spaniels (yes, again), and Labrador retrievers. These dogs have taken very different routes to capturing the hearts of American pet owners. The poodle's path to popularity was meteoric, but Labs took the slow and steady road to Number 1, increasing at a rate of about 10% a year over four decades. The graphs for cocker spaniels and beagles are wavy.

The graphs also revealed that most dog breeds never weasel their ways into our lives. How many of your friends own an otterhound, a breed that maxed out in 1993 at seventy puppies? Or a Harrier, whose registrations have hovered between six and forty since 1934? Soon I became obsessed with deciphering America's shifting canine landscape. At night, I dreamt about demographic curves. During the day, I spent way too much time flipping through breed growth charts, boring my friends with dog talk, thinking about obscure breeds with weird names—Keeshonds, Schipperkes, Puliks.

The fact is that I was getting nowhere.

HOW DOG BREEDS ARE LIKE BABY NAMES

My lucky break came one afternoon when I ran across an article in the journal *Biology Letters* written by two researchers I had never heard of. One was a biology graduate student at Duke University named Matt Hahn, the other Alex Bentley, a post-doc in anthropology at University College London. Their article was about names that people give their babies.

Matt and Alex were from different disciplines but they both studied how dumb luck affects evolutionary change—in Alex's case, how cultures change, and in Matt's, how genes evolve. Their hypothesis was that many shifts in human culture—from designs engraved on Neolithic pottery to country music hits—are attributable to the fact that humans are inveterate copycats. Baby names offered an ideal way to test their idea, as the Social Security Administration maintains a Web site with the frequencies of the 1,000 most common first names in the United States for each decade in the twentieth century.

Matt and Alex downloaded the names of a gazillion babies and went to work. Using computer models used to study changes in gene frequencies, they discovered that continual shifts in our favorite first names are explained by a theory developed to explain a mechanism of evolutionary change called random drift. The basic idea is startlingly simple: Styles change because people unconsciously copy one another. Occasionally, someone invents a new baby name or develops a new dog breed. Whether it gets copied and becomes popular is largely a matter of random chance. In matters of taste, we often follow the herd.

I admit that I could not follow the mathematics that Matt and Alex used to prove that baby names become popular by dumb luck, but I got the gist of their argument. It dawned on me that shifts in trendy puppies might offer another way to test their ideas. I emailed the authors and attached a couple of my dog graphs. For good measure, I added, "By the way, my data set consists of 48 million puppies. Interested?"

Their response was immediate and enthusiastic. "Yes! Send us the file." Their analysis revealed that just like baby names, dog breeds usually get popular by a throw of the cosmic dice. Our transient preferences for dog breeds follow a type of statistical distribution that mathematicians call *power laws*. In human societies, these show up in situations in which large numbers of people influence each other. Power law graphs are elegant in the way that a Brancusi statuette is. The line swoops down from the top left, dwindling gradually to right. Malcolm Gladwell, the *New Yorker* writer who has made a career of pointing out the real-world implications of arcane findings in the behavioral sciences, compares power law graphs to a hockey stick lying on the floor, blade pointing up.

The message of the power law is that, whether we are talking bestselling books, citations of scientific papers, music downloads, Web page hits, baby names, or dog breeds, roughly 20% of the available options will attract 80% of the attention, a phenomenon economists call the 80:20 rule. After the first couple of choices, popularity nosedives, dribbling closer and closer to zero. That's why in business circles, power laws have come to be referred to as "the long tail."

Here's how it works with dogs. In 2007, 81% of puppy registrations were from the top thirty-one breeds. This left the other 125 breeds to fight over the popularity crumbs. The bottom fifty breeds together attracted only 1% of all new puppy registrations. Labrador retrievers, the most popular breed, generated 9,000 times more puppy registrations than English fox hounds, the least popular breed. That's the inevitability of the long tail.

According to the random drift hypothesis, fashions are ever-changing because new innovations pop up every now and then. For example, a man invents an ugly plastic slip-on shoe he names after crocodiles, or someone develops a new dog breed by crossing a miniature schnauzer with a Yorkie (official name: Snorkie). Most new ideas are flops, but every now and then one will catch on. One implication of this throw-of-the-dice view of cultural change is that, for all practical purposes, it is impossible to accurately predict the next big thing—no matter if it is a shoe style, a baby name, a hit song, or a dog breed.

Our analysis of AKC registrations revealed that our tastes in dogs reflect the same mob psychology that leads people to think that nose rings are sexy. I am arguing, of course, that temporary enthusiasms for breeds are spread by mental viruses that, like biological ones, sometimes erupt into epidemics. An epidemic, whether swine flu or a new dance, goes through three stages. The first is slow, steady growth. Once it hits the proverbial tipping point, it spreads like fire until the third inevitable stage: burnout.

The same pattern plays out in dogs. In the 1950s, Irish setter registrations hovered between 2,000 and 3,000 a year. In 1962 they reached the tipping point and the breed "went viral." Registrations took off, jumping to over 60,000 in 1974, an increase of 2,300%. Then, as suddenly as it had risen, their popularity plummeted. By the end of the burnout stage, Irish setter registrations were only 5% of what they were at the peak of their popularity. The graph of their rise and fall is perfectly symmetrical, and, from start to finish, their fifteen minutes of fame lasted exactly twenty-five years. A dozen other breeds, including Dobermans, Old English sheepdogs, and Saint Bernards, have shown the same pattern.

THE RISE AND FALL OF THE PEDIGREED DOG

Epidemics in the popularity of these breeds illustrate the influence that cultural whims can have on our relationships with animals. Presently, Americans seem to be in the burnout phase of another canine fad—the desire to own a purebred dog.

The transformation of American dogs from pals and working partners to fashion statements began on a warm September day in 1884, when a group of sportsmen met in Philadelphia to form what became the American Kennel Club. They were following the lead of their British counterparts who, a decade earlier, had established the first all-breed kennel club, called, appropriately, the Kennel Club. The AKC soon established a registry that originally included 1,400 pedigree dogs representing eight breeds. As in England, the growth of enthusiasm for purebreds in the United States was spectacular. Dog shows began as a pastime of the landed gentry, but in the second half of the nineteenth century, the dog fancy caught the interest of the growing middle class. It was a classic case of fashion trickle-down from the rich to the wanna-be-rich.

Between 1900 and 1939, annual AKC registrations went from 5,000 puppies to 80,000. The big craze for purebreds came after World War II, when the proportion of American dogs that were purebreds jumped from 5% to 50%, and registrations were growing fifteen times faster than the human population of the United States.

I trace the explosion in pedigree dogs to the G.I. Bill of 1944, which enabled millions of Americans to buy homes in the suburbs with dog-friendly yards. My family was typical. My father, a bomber pilot, came home from the war in 1945. With a low-interest loan from the Veterans Administration, my parents bought a house with a lawn and got a beagle for my sister and me to play with. We were proud that Bosco had "papers" from the AKC. In truth, no one in our family really knew what AKC registration meant, but it sounded good to us. Our puppy was an aristocrat.

By 1970, the AKC was processing a million new registrations every year and was the world's largest registry of purebred dogs. The '80s and '90s were flush times for the organization. In 1990, one half of eligible dogs in the United States were registered with the AKC. In 2007, the AKC, a not-for-profit organization, took in $72 million, about half of which was generated by puppy registration fees.

Dark clouds, however, loomed. Long controversial, the AKC took a major hit in 1990 with the publication of a devastating critique in the

Atlantic Monthly. The author, Mark Derr, depicted the AKC as an elitist and secretive organization that was myopically focused on profits derived from the overbreeding of dogs for good looks. In 1993, registrations began a decline that looked increasingly like a death spiral.

Over the last fifteen years, AKC registrations have plummeted 50% from their mid-1990s peak of a million and a half new puppies each year. In 2008, the board of delegates was informed that if present trends continued, registrations would soon drop to 250,000 puppies, and the AKC would face a $40 million shortfall. In his report to the board, chairman Ron Menaker did not mince words: "Make no mistake, the very future of the AKC and the sport as we know it is at risk." He warned delegates that if the hemorrhage in registrations is not stemmed, the AKC will join the ranks of extinct corporate icons like Westinghouse, Pan American Airlines, and the Standard Oil Company.

The AKC is falling victim to changing cultural attitudes about the desirability of genetic purity in dogs. An AKC conformation show is a curious mix of eugenics and philosophy in which breeders chase an elusive Platonic ideal. Show dog people tell you that their sport is ultimately about improving the breed, bringing it closer and closer to perfection. Perfection in the show world is defined by a written code—the breed standard. According to the AKC standard, a Yorkie with a one-inch white spot on its chest is OK. A Yorkie with a two-inch white spot is disqualified. My favorite rule dictates that a Clumber spaniel have a "pensive expression."

While the breed standards pay lip service to a dog's temperament, in reality there is more emphasis on the color of the rims of its eyes and shape of its head than on traits that would make it fun to live with. I asked a handler at the Asheville show how much it would cost me to purchase a show-quality Silky Terrier. He said $2,000 to $3,000, but then added that he could put me in touch with a breeder who would sell me a "pet quality" Silky for around $800. In the dog show world, "pet quality" means "loser."

But a good pet is exactly what most people are looking for in a dog. In other words, the animals that breeders are trying to produce are different from the animals that most people want. And, in their efforts to create the perfect dog, professional breeders have produced an animal whose elegant beauty is literally skin deep—a dog that looks terrific until you peer under the hood. That is the paradox of the show ring.

Mary Jean and I are suckers for big, good-natured dogs, dogs you can trust, dogs that like children. That's why we are drawn to Labs and golden retrievers. We have had three, and we have loved all of them. In each case, however, the hereditary burdens of their tribe came bundled with their registration papers. Both Molly and Tsali, our Labs, developed crippling hip dysplasia. In Tsali's case, it was so bad that we had her euthanized when getting up in the morning had become an exercise in pain management. Dixie inherited a full complement of golden retriever maladies—dermatitis, hip problems, hypothyroidism, and congestive heart failure, the condition that finally did her in. Hereditary diseases are the rule, not the exception, among purebreds. Over 350 disorders lurk among the 19,000 genes that dogs carry around. Because many of these are shared with humans, purebred dogs have become a favorite animal model for the study of human diseases like narcolepsy, epilepsy, and cancer.

Pedigree dogs are especially susceptible to genetic disorders for several reasons. Some are the consequence of intentional selection for physical deformity. The best example is the bulldog. To preserve the breed, bulldog owners sought to transform an animal designed to latch on to the nose of a raging bull into a household pet. They did this by selecting for docility rather than athleticism and pugnacity. Bulldogs with monstrous heads and pushed-in faces became fashionable. This look was achieved by selecting for a skeletal malformation called chondrodystrophy. These distortions of the head and face produced a host of problems, including the inability to deliver puppies through the birth canal, labored breathing, snoring, and sleep apnea. Similar artificial selection for exaggerated structural features was responsible for bad hips in German shepherds (the consequence of breeding for sloping hindquarters) and bad backs in dachshunds.

Most genetic problems in pedigree dogs, however, are the by-product of inbreeding, not intentional selection for morphological weirdness. The majority of the 400 or so existing breeds of dogs were developed in the last 200 years from small pools of sires. All of the 20,000 AKC registered Portuguese water dogs trace their ancestry to thirty-one animals, and 90% of water dog genes come from only ten dogs. Inbreeding from a limited gene pool means that a puppy is more apt to inherit harmful recessive alleles from both its parents. Bad genes in just one ancestral line can do a lot of damage to a lot of dogs. This is known as the popular sire effect. Springer spaniels, for example, have the annoying habit of biting the hand that feeds them. Researchers at Cornell and the University of Pennsylvania have linked the occurrence of aggression in this breed to one kennel, and more specifically, to a single dog. (The researchers also found that lines of springer spaniels bred for the show ring are more apt to turn on their owners than are springers bred for hunting.)

When 25% of a product is defective, you would expect a consumer revolt. As the bloom on registered purebreds has faded, the mixed-breed dog, which until the 1950s was the most popular pet in America, has regained its cultural panache. In the 1980s, animal protection groups began to promote the adoption of dogs relinquished to animal shelters, regardless of their pedigree, as morally superior to purchasing a purebred puppy. In late 2008, Vice President Joe Biden's family picked a purebred German shepherd puppy to take with them to Washington. The headline on PETA's blog read "Joe Biden Buys One, Gets One Killed." The Bidens suddenly decided that they really needed a second dog, a rescue dog.

ARE WE RUNNING OUT OF GOOD DOGS?

As with other shifts in our tastes in dogs, the desirability of owning mixed-breed and rescue dogs has had an unanticipated consequence: As the number of people who want to adopt shelter dogs has gone up, the supply has gone down—way down. In 1970, 23 million dogs and cats were euthanized in animal shelters in the United States. In 2007, the number had dropped to an estimated 4 million. This drastic reduction in the number of abandoned pets is the result of the most successful campaign in the history of the modern animal protection movement: the spay and neuter crusade.

One result of our rush to pluck the gonads from every household pet is that America may be running out of dogs. Richard Avanzino, president of Maddie's Fund, a $300 million foundation that aims to eliminate euthanasia in animal shelters, worries that the slack will be taken up by puppy farms in China and Mexico. Others in the animal rescue community are not so sure that there is a dog shortage. They argue that we have plenty of adoptable dogs, they are just in the wrong parts of the country.

By some estimates, as many as 90% of adoptable dogs in some parts of the Northeast are ex-pats from the South. My county exports 200 shelter dogs north every year, and the Atlanta Humane Society packs off 600 of them. Rescue Waggin', part of PetSmart Charities, moves about 10,000 dogs a year from parts of the country where they aren't wanted to places where they are needed. The abundance of unwanted dogs in southern states reflects regional differences in attitudes toward neutering pets. Spay and neuter campaigns have been much more successful in the Northeast, the Midwest, and on the West Coast than in my part of the country. The per capita euthanasia rate of unwanted pets is forty times higher in North Carolina, for example, than it is in Connecticut. Our local animal rescue group tried to prod the town council into enacting a mandatory spay and neuter law. It was a nonstarter. People in the rural South don't like restrictions on their dogs any more than they like zoning, gun control, or laws that keep you from carting little children around town in the back of an open pickup truck.

My friend Jill's day job is teaching history to college freshmen, but her passion is finding homes for abandoned dogs. Over the years she has saved 2,000 dogs from the needle. Once a month, she plays God. It's heartbreaking. Jill walks the aisles of our local shelter carrying an arm made out of rubber. If she thinks a dog has a future, that someone could fall in love with it, she runs the dog through a standardized behavioral screening test. That's what the fake arm is for. She puts a bowl of food right in front of the dog's nose. Then, once it starts eating, she snatches the food away with the rubber hand. If the dog growls or goes for the hand, it is toast.

Most dogs don't make the cut. Some of them bite the hand. Others are unadoptable because they are old or ugly. Because small dogs are "in," it has gotten harder to place big dogs, especially the black ones. Jill also avoids pit bulls. Too often, she says, they are adopted by the wrong kind of people. Almost all the animals she rejects will be euthanized. The dogs she deems adoptable will be vaccinated against rabies and parvo. Two weeks later, they will join two dozen other dogs in the back of a truck headed up I-95, riding the underground dog railway to suburbs in the Northeast.

Not all of Jill's migrants work out. Last year, she got a call from an animal shelter near Greenwich, Connecticut, that was not happy with one of the dogs Jill had sent them, an Australian shepherd mix. He did OK on Jill's rubber hand test, but he got seriously cranky once the truck passed the Mason-Dixon Line. The shelter gave her a choice: Either you come and get the dog, or we euthanize it. Jill got in her car that evening, drove fourteen hours up the interstate, picked up the dog, turned around, and drove fourteen hours home. It was a wasted effort. The Yankees were right; the dog was dangerous. A month later, Jill had to put him down.

THE HUMAN-DOG RELATIONSHIP IS A MIXED BAG

The decision of a handful of wolves 20,000 or 30,000 years ago to entrust the destiny of their species to humans has been a mixed bag. Over time, the descendents of these animals weaseled their way into our hearts and the fabric of human society. The upside was a steady food supply and a warm fire to curl up by. The fact that the dogs' bipedal benefactors cooked a puppy now and then seemed a small price to pay for the security of home and hearth. But then came the leash, the collar, the fighting pit, obedience school, the animal shelter. Finally, there was the ignominy of genetic sculpting, in which humans transformed the basic canine form—a model of grace and efficiency—into shapes, shades, and sizes never found in nature. We have jiggered the dog genome into a bewildering array of animals that look magnificent but that, in the end, are like modern tomatoes, a triumph of style over substance. We now choose our animal companions using the same facile consumer psychology that we apply to choosing the latest clothing styles. The transition from function to friend to fashion statement is complete.

Yet, at the same time, we love our dogs. We confide in them, buy them Christmas presents, take them on vacations, make Web sites for them, and treat them like our children. They sleep in our beds. We grieve when they die. The paradoxes of the human-dog relationship are particularly evident in people who are the most devoted to their animals. I admire the passion of purebred dog fanciers whose lives are focused on improving their breed. Their love for dogs is deep and real. But in their efforts to create the perfect dog, they have produced millions of animals with itchy skin, skulls that are too big, hearts that are too small, and hips that always hurt. Animals that suffer.

Nancy Brown's mission in life is to make her posse of wolf-dogs as happy as possible. But the hard truth is that these creatures are an uneasy mix of the wild and the tame, destined to spend their lives in the confines of a chain-link pen.

Jill knows that two dogs will be euthanized in my county for every one that she finds a home for. She bears the moral burden of deciding which ones will live and which ones will die. Her commitment to dogs that no one else cares about has restricted her social life and interfered with her career. She looks sad most of the time. She got it right when she told me, "This is not something that you can do a little of."

My own relationship with dogs is also complicated. I have loved all the dogs we have lived with, even Puppy, a Benji look-alike we rescued six years ago. She was the smartest dog I have ever met. Puppy's innocent face, however, belied a mean streak that got worse as she got older. She terrorized our aging Lab and eventually bit every member of the family, some several times. I consulted some of the best animal behaviorists in the country to no avail. One afternoon, with no warning, Puppy attacked a friend who was visiting. It was the last straw. Dr. Shields, our veterinarian, told me that I needed to face reality. It was time to put her down.

I remember that last long drive to the vet's office as if it were yesterday. It's been four years, but her death still weighs heavily on me. Maybe that's why we have a cat now. But I do miss having a dog around the house.

5

"Prom Queen Kills First Deer on Sixteenth Birthday"

GENDER AND THE HUMAN-ANIMAL RELATIONSHIP

> As a society, we expect women to respond to the suffering of animals; we see that as a "natural" part of womanhood.   —BRIAN LUKE

> Men enjoy hunting and killing, and these activities are continued in sports even when they are no longer economically necessary.
>   —SHERWOOD WASHBURN

When I began my first forays into anthrozoology, I was struck by how differently men and women interacted with animals. For instance, I once interviewed veterinary school students about their reactions to moral issues such as euthanizing healthy animals. One of the women, Elizabeth, told me, "I cried the first time and I cried the fifteenth time. It hasn't gotten any easier, but I have learned to mask my feelings in front of the client, to be strong for them." Her male counterpart, William, had a completely different response. "Euthanasia doesn't affect me," he said. "I wonder about myself sometimes, but, honestly, I don't feel bothered by it."

After hearing a lot of similar statements, I came up with an obvious hypothesis; namely, women are nice to animals and men, well . . . not so much. Then I began to stumble on anomalies, the people who don't fit our stereotypes about how the sexes relate to other species—women like Evelyn Clancy and men like Bill Gibson.

WHEN STEREOTYPES FAIL: STUDYING
THE GENDER BENDERS

One afternoon, a graduate student named Amy Early walked into my office and told me she wanted to write her thesis on women hunters. It was good timing. I had just read a story in my town newspaper about a young woman who, on her sixteenth birthday, put a .30-30 slug in the chest of an eight-point buck. In addition to being an ace hunter, she was also the high school prom queen. The newspaper story made me wonder how women who hunt negotiate the delicate line between nurturance and their desire to kill animals.

My eyes lit up when Amy told me about her thesis idea. "Great project. Let's run with it."

One of the first people Amy interviewed was Evelyn Clancy , a middle-aged housewife who loved animals but who also loved to shoot them. Her passion was big game. She and her husband had been on several African safaris, but she had never bagged a zebra. One afternoon, Evelyn and her hunting guide, Anthony, came upon a large zebra herd. Whispering, she asked Anthony which of the animals she should aim for. He pointed out a big male. She snapped a bullet into the chamber, got the animal in the crosshairs of the scope, took a breath, held it, and squeezed the trigger. The zebra dropped.

Anthony turned and said, "That's not the right one." They walked over to the zebra. It was a female, and Evelyn burst into tears.

"I had shot a female," she told Amy. "I cried all night. I had tears in my eyes the next morning. It was the only time I ever shot the wrong animal. I felt awful about killing her. She may have been a mother. She may have had a little colt running around. She could be pregnant. That's the reason it upset me so much, shooting the wrong zebra."

Evelyn's story says a lot about the complexity of gender in our interactions with other species. Her excitement at bagging her first zebra turned to despair the moment she realized she had killed a creature with whom she identified. How many male hunters would cry all night because they had shot the wrong animal? And if they did, how many would admit it?

Bill Gibson represents a different type of gender bender. He is the skeptical, hard-nosed research psychologist who inhabits the office next to mine. One afternoon I asked him about the faded photograph of a dog he keeps on his desk. Bill is a math guy, a left-brainer. He designs computer systems that tell which part of your cortex is working overtime. But when he started talking about Blue, a shepherd-husky mix, he melted.

He told me that he named the dog Blue because he had one blue eye and one brown eye. He bought Blue back in '84 from a guy who ran a puppy mill. For eleven years, Blue was Bill's best friend. "Blue would look me straight in the eye. He knew things." Bill said.

He told me how Blue had suffered through cancer of the bowel, and the prognosis was bleak.

"I was with him at the end." Bill said. "I put a nice blanket on him and gave him a kiss good-bye and talked to him. I had him cremated. I kept the ashes for a while and then later spread them on a mountain. I was a mess. Sometimes I would break into tears. It was embarrassing."

Bill started to choke up. "I still think about him every day. I loved him more than I've loved any person in the world. I'll probably think of him on my death bed."

Evelyn and Bill illustrate the difficulty of making simple generalizations about men, women, and animals. Cases like these convinced me that my initial hypothesis about sex roles and how we treat animals was wrong—clearly, some men are more sensitive to the plight of animals than some women. But then there are the many instances in which sex differences do fit gender stereotypes. For instance, all the cockfighters I knew were men, and a large majority of animal rights activists were women. I decided to take a more systematic approach to examining how gender affects our relationships with other species. I started amassing every study I could find on the topic. This quest took me from A (animal activism) to Z (zoophilia). I discovered that some aspects of our relationships with animals are profoundly affected by gender. But I also found that both the direction of sex differences and their size depend on the type of relationship we are talking about.

WHICH SEX LOVES PETS THE MOST?

Surprisingly, gender differences in attachment to pets are smaller than most people think. In the United States, equal numbers of men and women own companion animals, they are just as likely to buy holiday presents for their dogs and cats and to pay newspapers to publish obituaries for their deceased pets. The sexes are even similar in whether they let their animals sleep in their bed. (Women get the nod, but not by much.) But are men as bonded to their pets as women are?

Anthrozoologists have developed standardized questionnaires to assess how much individuals love their pets. For example, one widely used scale asks people how much they agree or disagree with statements such as "I would do almost anything to take care of my pet" and "My pet means more to me than any of my friends." When it comes to pet love, women seem to have a slight edge over men, but the difference is surprisingly small. I examined a dozen studies that reported sex differences in attachment to pets. True, women scored higher than men in most of them, but the difference between the attachment score of the average man and that of the average woman was usually negligible.

In some aspects of caring for pets, however, big gender differences do show up. You will not be surprised to learn that women are twice as likely as men to dress their pets in little outfits. And, as is the case in many spheres of human life, women do more than their fair share of pet-related chores. In three out of four American homes, women are the ones who usually feed the family dog and clean out the cat's litter box. Women also make up about 85% of veterinary clients. (Several veterinarians told me that when men bring pets into their clinics, they often carry notes from their wives spelling out exactly what is ailing Spot or Fluffy.)

I have always assumed that males and females play with pets differently. After all, boys are more apt to engage in the mock wrestling and hand-to-hand combat that developmental psychologists call *rough-and-tumble play*. In my house, there is gender bias on how we relate to our family pets. For example, Mary Jean and our twin daughters, Betsy and Katie, would gently pat our dogs on the head while Adam and I were more apt to wrestle and play chase and pull-and-tug with them.

There is, however, little evidence that my hypothesis applies to anyone outside the Herzog household. Italian researchers found that that men and women were not any different in how they played with their dogs, and a study of the interactions between dogs and their owners in veterinary hospital waiting rooms also found no differences in the behavior of men and women. Gail Melson, a developmental psychologist, asked parents to estimate how much interest their children had in playing with younger children, babies, stuffed animals, baby dolls, and pets. As you would expect, girls were much more interested than boys in babies, dolls, and stuffed animals. But, to my surprise, she found no sex differences in how kids play with or nurture their pets. Gail believes that for boys, pets are often the only vehicles that give them experience in caring for another living being.

Women, however, are more susceptible than men to creatures that are cute. British researchers recently reported that two groups of women are particularly sensitive to differences in the cuteness of infants: those of reproductive age and those taking birth control pills that raise their levels of the hormones progesterone and estrogen. Cute animals have the same effect on women. University of California at Santa Barbara researchers were interested in changes in the attractiveness of a golden retriever puppy named Goldie as she matured. Over a five-month period, they took Goldie to a highly traveled spot on campus where she would sit for an hour with her "owner" (actually an assistant in the study), while the researchers tallied the number of passersby who came over to pet or play with her.

Goldie's ability to seduce strangers decreased precipitously as she transitioned from puppy to adult. Her drop in popularity was especially steep among women. When Goldie was at her cutest, women were twice as likely as men to chat her up. But by the end of the study, the number of women who stopped to stroke her head and say hi had dropped 95%, and the sex difference had completely disappeared.

SEX DIFFERENCES IN ATTITUDES TOWARD ANIMALS

Stephen Kellert of Yale University has spent his career investigating the peculiarities of human attitudes toward other species. He has consistently found that women are more concerned with protecting animals than men are. But, at the same time, phobias of creatures like snakes and spiders are three times more common in women than men. And while men know more about the biology and ecology of other species than women, they tend to appreciate animals for what Kellert calls "practical and recreational reasons." In other words, men are more likely to approve of killing animals for fun and profit.

These gender differences in attitudes toward animals also exist in other societies. Linda Pifer of the Chicago Academy of Sciences asked adults in the United States, Japan, and thirteen European countries how they felt about using dogs and chimpanzees in experiments aimed at developing treatments for human afflictions. In every country, women were more opposed to animal research than men. Swedish researchers reviewed dozens of studies conducted worldwide and did not find a single country in which more women than men supported animal research.

Glib generalizations about sex differences and animal attitudes, however, can be misleading. In attitudes about the use of animals, men and women are more similar than they are different. For example, the National Opinion Research Center asked a large sample of men and women how they felt about this statement: *It is right to use animals for medical testing if it might save human lives.* They found that more men than women "strongly agreed" with the statement and more women "strongly disagreed." However, most people were in the middle on this issue, and some women were more supportive of animal experimentation than some of the men. This finding illustrates one of the most important facts about human gender differences. It is that in nearly every human psychological characteristic, men and women overlap. This means that in most cases the differences within the sexes are bigger than the differences between the sexes.

WOMEN TAKE ACTION!

But while men and women are fairly similar when it comes to *attitudes* about animal welfare, the sexes are not at all alike when it comes to *taking action* on behalf of other species.

On September 12, 2005, anthrozoologist Leslie Irvine and three of her friends hopped on a plane in Denver and headed for Louisiana. Their destination was the Lamar-Dixon Expo Center in Gonzales, the main staging area for animal rescue operations in the New Orleans area after Hurricane Katrina. The group was unprepared for the chaos they encountered. That night, Leslie wrote in her field notes, "Who can imagine the sound of a thousand dogs barking? Until today, the question would have seemed like a perverse *koan*. But now that I know what a thousand dogs sound like, I wish everyone could hear. It sounds like futility, helplessness, and the desperation of this undertaking. . . . I am sure it will haunt me for a very long time." Six days later, Leslie became a Katrina victim herself. Mentally and physically exhausted by nearly a week of animal rescue, she collapsed and was taken to the hospital. Three years later, she was still haunted by the sound of a thousand dogs barking. There were lots of heroes like Leslie who saved (or tried to save) the lost animals of Katrina, and the vast majority were women.

The modern animal rights movement has recruited women in droves; every study of the sociology of animal liberation has found that three to four times as many women as men boycott circuses, march against animal experimentation, and drive cars plastered with "Meat Is Murder" bumper stickers. Interestingly, the sex ratio of individuals involved in animal protection has not changed at all over the last 150 years. Even in the Victorian era, four out of five members of animal welfare organizations were female.

The predominance of women among people who take animals seriously is not limited to animal liberators. At the National Zoo in Washington, D.C., hundreds of dedicated volunteers, most of them women, stand next to exhibits for hours on end, patiently explaining a hundred times a day that the alligator snapping turtle really is alive despite the fact that it looks like a hundred-pound hunk of moss-covered granite, that Ambika the elephant is sixty years old, and that a pygmy hippo is not the same as a baby hippo. You also see big sex differences among the keepers. As I strolled around the grounds one snowy day in January, I ran into a zoo employee. She told me that that every one of the keepers who cared for the chimpanzees, gorillas, and orangutans in the Great Ape House was female.

Women dominate nearly every aspect of grassroots animal protection. They make up 85% of the membership of the two largest mainstream animal protectionist organizations in the United States, the ASPCA and the Humane Society of the United States. Among dog rescuers, women outnumber men eleven to one, and three times more female high school students than males call the National Anti-Vivisection Society's dissection hotline each year because they want to opt out of biology dissection labs for reason of conscience. And more women than men give up meat for ethical reasons.

THE DARK SIDE OF THE BOND: SEX DIFFERENCES IN ANIMAL CRUELTY

But don't make the mistake of thinking that all women are nice to animals.

One Thursday morning in January 2006, Joanne Hinojosa was arguing with her estranged husband in front of his home in South Austin, Texas.

At some point in their spat, things took a turn for the worse. She took a swing at him and he ran down the street to call the cops. That's when Joanne went after his dog. She carried the twenty-pound mixed-breed female named Marti into the house and started stabbing her. The police found Marti lying in a pool of blood with a knife sticking out of her left side. The dog had been stabbed twenty-seven times. They rushed her to the Ben White Pet Hospital for emergency surgery, but it was too late. Hinojosa's lawyer claimed his client suffered from post-traumatic stress disorder. She pled guilty to animal cruelty and was sentenced to six months in jail and anger management training.

Spousal animal abuse is surprisingly common. Pets are often caught up in domestic violence disputes. Developmental psychologist Frank Ascione, executive director of the Institute for Human-Animal Connection at the University of Denver, found that over 70% of battered women he has studied said that their partners had abused, threatened to abuse, or had killed a pet. Their violence ranged from shooting the family dog to setting the children's kitten on fire. What is unusual about the Hinojosa case is that the perpetrator was a woman.

The data at Pet-abuse.com, a Web site that tracks media reports of animal cruelty cases, offers a window into sex differences in animal cruelty. An analysis of the 15,000 cases in the site's database reveals that men are involved in 70% of these incidents. But this figure is misleading. The sexes are not very different in the frequency with which they are charged with animal neglect, which is usually the result of apathy, poor judgment, or stupidity rather than malicious cruelty. If, however, we omit these incidents, it becomes clear that men commit nearly all of the really nasty offenses against animals: 94% of beatings, 91% of burnings, 84% of stranglings, 94% of hangings, 92% of stomping to death, 94% of shootings, and 95% of stabbings.

PEOPLE WHO LOVE ANIMALS TOO MUCH:
GENDER AND ANIMAL HOARDING

There is one exception to the rule that most animal abusers are men. Three times as many women as men get caught up in animal hoarding. One semester I invited a woman who was an ordained minister to be a guest speaker in a class I was teaching. After her lecture, she mentioned to me that her house was for sale. I perked up as Mary Jean and I were in the market for a new home. I drove out to look at the house the next day. It was halfway up the side of a mountain near Greens Creek. The property had privacy, a southern exposure, and a big view toward north Georgia. Just what we were looking for. I had my hopes up. Then I opened the front door.

"Overpowering" does not quite capture the stench of feces and urine. Cats and clothing were strewn about the living room. A forty-pound bag of dried cat chow lay spilled open in the middle of the floor. There was no place to sit. The woman, who had seemed perfectly normal when I talked to her the day before, was in a tizzy. She apologized to me, saying that things had gotten a bit out of hand, that she had tried to straighten things up for my visit. I counted two dozen cats, but there were probably others roaming the nearby woods. I did not stay long, and we did not buy the house. We could have probably gotten it cheap.

The Hoarding of Animals Research Consortium, an interdisciplinary group of researchers led by Gary Patronek estimates that up to 2,000 cases of hoarding involving 200,000 animals are reported in the United States each year. People usually don't think that hoarding is as cruel as beating and shooting animals, but from the animals' perspective, a hoarding situation can be worse; the suffering can go on for years.

The stereotype of an animal hoarder is the crazy old lady living alone whose neighbors call the health department when they finally tire of the noise and the stink. Is this image accurate? Sometimes it is. Seventy-five to 85% of hoarders are women. Half live alone, and half are over sixty.

According to Arnie Arluke and Celest Killen, authors of the book *Inside Hoarding: The Case of Barbara Erickson and Her 552 Dogs,* the woman whose house I saw would probably fit into the *incipient hoarder* category. With twenty or thirty animals living in her house, her life was just starting to fall apart. The numbers of animals confiscated from a hoarder's home can be unimaginable. The record is the "Great Bunny Rescue of 2006," in which nearly 1,700 rabbits were rescued from the backyard of a woman named Jackie Decker who lived in a two-bedroom house near Reno, Nevada.

A recent study on the public health implications of animal hoarding, reported that nearly all hoarders who have over 100 animals in their homes were women. The living conditions of these extreme hoarders ranged from lousy to horrifying. Things were particularly bad among those living alone. Over half of their houses lacked stoves, hot running water, or working sinks and toilets. Forty percent of the homes had no heat and 80% did not contain a functional shower or refrigerator. The conditions for the animals living in these circumstances are dire—cats, dogs, pot-bellied pigs, rabbits, all emaciated and ridden with disease, all running amok. First responders called in to clean up hoarding situations often encounter half-eaten animals corpses lying about.

Clinicians have come up with several explanations of why people hoard animals. The wildest theory is that cat hoarding could be caused by toxoplasmosis infection. The idea that hoarding is the result of a parasite that rewires your brain remains unproven, but I find it intriguing. Tox-infected rodents suddenly become attracted to cats and can't seem to get enough of the smell of cat urine. In humans, tox infection has been associated with mental illness and neuroticism. It is not a far stretch to imagine that a brain parasite might cloud the judgment of people with lots of cats and perhaps even zap out enough neurons so that they become inured to the smell of tomcat pee. The conventional theories, however, link hoarding to dementia, obsessive-compulsive disorder, addictive personality, defects in social attachments, or delusional thinking. Hoarders are usually convinced that their animals are happy and that the hoarder has a special ability to communicate with them.

While anthrozoologists don't know the exact cause of hoarding, researchers agree that is nearly impossible to cure the disorder. The recidivism rate among hoarders is nearly 100%. In the Reno bunny case, for instance, animal control officers had confiscated 500 rabbits from the same property just four years earlier. Despite the fact that most of these animals had to be euthanized, a local judge treated the case like a joke and turned down a request for a court order that would have kept the hoarder from obtaining even more animals.

I suspect that hoarding sometimes results from a perfect storm of the well-intended desire to save animals and the inability to draw lines in the moral sand. This means that individuals drawn to animal rescue, most of whom are women, are at special risk. For the past couple of years, my students and I have been interviewing members of animal rescue organizations. While the vast majority are perfectly normal, and some of them are saints, occasionally we run into red flags. During these interviews, we ask rescuers how many pets they have. Most people say something like, "two dogs, a cat, and parrot." But sometimes when we ask about their pets, a rescuer will look down, laugh a little, and say something like, "Oh, hmm . . . too many, I guess." These are the people who make it clear that they don't want us to interview them at their homes.

I got a twinkling of insight into how a person might not be able to just say no, how he or she might *have* to bring just one more animal into the house, when I was given a behind-the-scenes tour of a municipal animal shelter. I was immediately drawn to a young boxer lying on his side, panting, in a cage in the quarantine room. He lifted his head up, looked at me with the world's saddest eyes that pleaded, *Take me out of here, please?* I could feel my chest getting tight, and I was a goner. If he had not been contagious and unadoptable, I would have brought that dog home on the spot.

Eight thousand dogs and cats enter the shelter every year. Forty percent of them will leave with a new owner wearing a big grin. The other 60%, including most of the cats, and nearly all the large black dogs and pit bull mixes, will get a couple of cc's of sodium pentobarbital injected into the cephalic vein of their foreleg and slip away in a matter of seconds.

Becky, the director of the facility and my tour guide, has worked in shelters for fifteen years and she loves animals. She really does. As we squeeze past rows of stainless steel cages, she points to a Treeing Walker (a type of coon hound) and tells me that it was found running loose near Big Oak Gap, and then she calls an orange cat by its name. The back rooms are crowded and my ears ring from the incessant barking. It is too loud to talk easily. Some of the animals look fine, others are terrified.

Becky's passion is making life better for unwanted animals. She also has her own pets—three cats, three dogs, and three birds. The paradox of her profession becomes apparent when I ask, "How many dogs and cats have you euthanized over the years?"

She looks at me as if I am an idiot.

"Over a thousand?" I ask meekly.

After a pause, she says, "At least."

"How do you stay sane?"

"Somebody has to do it. I don't obsess about it," she says.

Then she shows me a message that appeared on a Listserv a couple of days ago. It was written by a shelter director who goes home every night and cries. The woman is bitter. "I hate my job," she wrote. "I hate that it exists and I hate that it will always be there unless you people make some changes and realize that the lives you are affecting go much further than the pets you dump at a shelter. I do my best to save every life I can, but there are more animals coming in every day than there are homes." Becky tells me that the other director is in the wrong line of work.

Remarkably, Becky is cheerful and passionate about her career. She is the right person for the job. This is not, however, true for all volunteers she supervises. She has to keep her eye on a few of them—the ones that have potential to be hoarders, the ones that don't have the moral strength to be animal rescuers.

THE NATURE (AND NURTURE) OF SEX DIFFERENCES
IN OUR INTERACTIONS WITH ANIMALS

After reading hundreds of articles about sex differences in our relationships with animals, I came to several conclusions. The first is that, as a
general rule, women have more of a soft spot for animals than men do.
The second is that while our stereotypes about the directions of gender
differences in relationships with animals are usually on the mark, our
beliefs about the size of the differences between men and women are
often wrong. There is not much difference at all between men and women
in the frequency with which they live with pets and in the way boys and
girls play with animals. Sex differences in attitudes of the general public
toward issues like animal research are a bit larger, but there is a lot of overlap between the sexes. The big sex differences only come into play when
you look at the extremes—at animal activists and animal abusers.

Everyone wants to know if human sex differences are a result of nature
or nurture. This is a loaded question. (Indeed, Lawrence Summers lost his
job as president of Harvard in part for suggesting that biology might play
a role in gender differences in scientific productivity.) It is also an intellectual loser. To think that complicated behaviors like the way humans
feel about animals is the result of either nature OR nurture is an example
of the myth of single causation. Lots of factors affect differences in how
the sexes relate to animals. Some social scientists argue that women are
drawn to animal rights because both women and animals are victims of
male exploitation and thus women identify with animals more than men
do. Others link sex differences to socialization. For example, in his book
Brutal: Manhood and the Exploitation of Animals, Brian Luke argues that
our culture instills indifference to animal suffering in boys practically
from birth.

Certainly some of the differences between men and women have their
origins in exploitation and socialization, but biology also plays a role in
human-animal relationships. For instance, hunting is defined as a male
activity in every human culture. Well, nearly every human culture. In the

Ituri Forest of the Democratic Republic of Congo, Bambuti pygmy women help drive game into nets, and Matses women in the Amazon Basin often accompany their husbands on hunting forays. They spot game and kill animals with spears and machetes. Because hunting trips offer privacy, a round of hot jungle sex can help the time go by when prey is scarce. For this reason, Matses men, who are polygamous, take only one of their wives on hunting expeditions, and women complain when they don't get invited to "go hunting" enough. Cultures like these, however, are exceedingly rare. In nearly all human societies, hunting expeditions are as sexually integrated as the Green Bay Packers locker room during half-time.

Developmental psychologists have found that some sex differences show up so early that they are unlikely to be the result of socialization. At the age of three months, boys outperform girls at tasks involving mentally rotating objects. And (I know this is hard to believe), several studies have now shown that male and female monkeys exhibit the same preferences in toys as human children. Male monkeys are attracted to "boy" toys (trucks, for example) and female monkeys like to play with soft, cuddly objects. Some differences in human reactions to animals are also found in infants. Baby girls, for instance, learn to associate fear with spiders and snakes more quickly than baby boys.

Our body chemistry also affects our interaction with other species. Several hormones influence empathy directed toward other people and animals. One of these is oxytocin, a chemical that switches on maternal instincts and facilitates social bonding. In humans, oxytocin levels rise during pregnancy and spike during childbirth, breastfeeding, and sexual orgasm. Oxytocin is also related to sex differences in empathy. Men, for example, are not as good as women at reading the emotions hidden in a face, but a whiff of oxytocin temporarily improves their emotional intelligence and makes men more generous.

Is oxytocin the glue that cements the human-animal bond? Meg Daley Olmert thinks so. In her book, *Made for Each Other: The Biology of the Human-Animal Bond,* she writes that pets are "fountains of oxytocin" and that pet lovers go through their day with an "oxytocin glow." Unfortunately, this claim is based on a single study that had only eighteen subjects. True, the researchers did find that oxytocin levels rose after people interacted with dogs—but they also found that oxytocin increased nearly as much when the subjects just sat quietly and read a book. More recent experiments on the role of oxytocin in our relationships with pets have reported mixed results. In one study, levels of the hormone increased in women who petted dogs, but they actually declined in men. A group of Japanese researchers found that whether or not oxytocin increased when owners played with their dogs depended on how much the dog gazed at its owner during the interactions. And a research team at the University of Missouri found that interacting with pets had no effect on their owner's oxytocin level. So, while oxytocin may be involved in some aspects of the human-animal bond, we need a lot more research before drawing conclusions about the role of this hormone in our attachment to pets.

The male hormone testosterone has the opposite effect on empathy. In both men and women, the more testosterone in the bloodstream, the more aggressive and less empathetic you will be. Testosterone also affects how people interact with their pets. Amanda Jones and Robert Josephs of the University of Texas found that in dog agility competitions, the amount of testosterone in a man's saliva affected how he treated his canine partner after the event. High-testosterone men whose animals did not perform well punished or even hit their dogs. Low-testosterone men, in contrast, lavished love on their animals no matter how they placed in the trials.

HOW BELL CURVES ACCOUNT FOR HUMAN SEX DIFFERENCES IN OUR RELATIONSHIPS WITH ANIMALS

The bottom line is that sex differences in our interactions with other species are the result of an inextricable mix of political, cultural, evolutionary, and even biochemical forces. Is there a way to make sense of sex differences in how we think about animals without resorting to clichés in the debate over nature and nurture? Yes.

Some years ago, I came across a little-known article in the *New Yorker* by *Tipping Point* author Malcolm Gladwell. The title was "The Sports Taboo: Why Blacks Are Like Boys and Whites Are Like Girls." In it, Gladwell argued that to really get a handle on racial and gender differences, you need to understand the bell-shaped curves statisticians call "normal distributions." Bell curves describe many psychological and biological phenomena. The basic idea is simple. For traits ranging from extroversion to the size of goldfinch beaks, most cases will fall near the middle of the pack, with the numbers tapering off as you move toward the extremes of the distribution. IQ test scores are a good example of a bell-curve trait. The average IQ score in the United States is 100. While 50% of people have IQs greater than 100, only 2% score over 130 on IQ tests, and one person in 1,000 gets above 145.

Bell-curve thinking is sometimes—and wrongly—thought of as racist. That's because in a 1994 book called simply *The Bell Curve*, Richard Herrnstein, a psychologist, and Charles Murray, a political scientist, used normal distributions to support their contention that racial differences in IQ are inherited. But bell curves are just shapes. They say absolutely nothing about whether differences between groups are due to genes or environment, or both. While bell curves do not explain the ultimate causes of sex differences, they can help us understand why most animal activists are women and most animal abusers are men.

My position is that many human gender differences, including how we treat animals, are simply the consequence of an elegant statistical principle that even most psychologists don't understand. It is this: *When two bell curves overlap, even a small difference between the average scores of the groups will produce big differences at the extremes.*

Height is a good example. In the United Sates, the average man is 8% taller than the average woman. This does not sound like much, but the sex ratio gets more and more out of whack as we go toward the high and the low ends. For example, among people over five foot ten inches, there are thirty men for every woman, but the sex ratio shoots up to 2,000:1 when we look at people over six feet.

The statistical principle that small differences between the average male and average female make for big sex differences at the extremes explains sex differences in many areas of human behavior. The fact that women are ten times as likely as men to die from the complications of anorexia follows directly from the fact that the average American woman is a bit more concerned with body image than the average man. And the enormous sex difference in homicide rates is a consequence of the real but surprisingly small difference in the aggressive tendencies of the average man and woman.

Here is how the Herzog (stolen from Malcolm Gladwell) theory of sex differences plays out in human-animal relationships. Assume for a moment that Americans vary in a hypothetical psychological trait called "liking pets," and that the distribution of this trait is bell-shaped—most people are in the middle, but a small proportion of people pathologically dote on their animals and a few people truly hate animals. Assume also that the average woman scores slightly higher than the average man on this trait but, as is nearly always the case, that there is a lot of overlap between the sexes. If my bell-curve idea is correct, as we move toward the pro-pet and anti-pet extremes, bigger and bigger gender differences should emerge.

This, of course, is exactly what we find. Among extreme pet lovers (off-the-chart hoarders), women outnumber men ten to one; and among serious pet haters (sadistic animal abusers), the male-to-female ratio is even more skewed. Overlapping bell curves also explain why so many animal rights activists are women. Public opinion polls indicate that as a group, women are more concerned with the welfare of animals than men are. The difference, however, is not all that big, and the magnitude of differences in attitudes toward animals within the sexes is a lot bigger than the differences between the average male and the average female. But, once again, as we move toward the tails, the sexes go their different ways. On the pro-animal side, four times as many women as men donate money to the ASPCA, boycott circuses, and show up at animal rights demonstrations. On the anti-animal side, vastly more men than women find pleasure in shooting animals for sport.

The bell curve explains a wide array of sex differences in human-animal interactions. It works regardless of whether evolution or culture is responsible for the six-year-old girl's inner voice the first time she visits the zoo—*I feel so sorry for that cute little monkey in the cage*—or the inner voice of a teenage boy on his first father-son hunting expedition that says, *Exhale slowly. Hold. Pull the trigger . . . NOW.*

6

In the Eyes of the Beholder

THE COMPARATIVE CRUELTY OF COCKFIGHTS
AND HAPPY MEALS

> The people who set one animal against another haven't got the guts to be
> bullies themselves. They're just secondhand cowards.
>
> —CLEVELAND AMORY

> Cockfighting is the most humane, perhaps the only humane, sport there is.
>
> —CAPTAIN L. FITZ-BARNARD

I am driving toward Knoxville on Interstate 40 to interview Eddy Buckner, a cockfighter I hung out with years ago while I was writing my doctoral dissertation on the behavior of chickens and the psychology of chicken fighters. Ten miles from the Tennessee line, I notice white feathers whizzing past my windshield. I speed up, pass a couple of eighteen wheelers, and find myself behind two flatbed trucks, each loaded with thirty-four wire crates packed with live chickens, headed for the slaughterhouse. The crates are about three feet wide, four feet long, and ten inches high, and look as if they hold thirty or forty chickens each. I do the math in my head:

At three birds per square foot of cage space, that's over 1,000 animals on each truck, the chickens crammed together like anchovies packed in a jar of oil. It is fifty-five degrees outside and the chickens are exposed to the wind and highway noise. The trucks are doing eighty as we cross the state line into the aptly named Cocke County. The chickens are shivering and hiding their heads under their wings, terrified, feathers flying. I think to myself, *Why don't federal animal welfare laws cover the interstate transport of commercial poultry?*

I tail the trucks for twenty miles, past the Wilton Springs exit, only a stone's throw from the abandoned Four-Forty Cockpit, which closed down in 2005 when the FBI started a full-court press on east Tennessee rooster fighters. While a lot of chickens have been cut down in the Four-Forty main pit over the last thirty years, thousands of times more have died to feed the hordes of hungry Smoky Mountain tourists who will stop at the McDonald's just down the road for a six-piece McNugget Happy Meal. And it dawned on me that the comparative ethics of fighting chickens versus eating them is more complicated than most people realize.

If Molly, our Labrador retriever, had not developed a taste for raw eggs, I would never have discovered the subterranean world of illicit cockfighting that existed in my little community. Shortly after we moved to the mountains, Molly turned thief, and took to stealing eggs out of our neighbor Hobart's henhouse. I got an inkling that something was amiss when I drove home one afternoon and found Molly lying on the porch with an injured leg and egg, literally, on her face. The next day, Mary Jean ran into Hobart's wife, Laney, at the grocery store, and mentioned that Molly had been hurt, but we did not know how. "Oh," Laney said. "Hobart caught her in our henhouse again and shot her in the rear end. Don't worry none. He just peppered her with birdshot."

I was not mad at Hobart. After all, my dog was stealing his eggs. But I knew I had to do something fast. The solution, I figured, was to get my own chickens and teach Molly not to mess with them. I looked in the classifieds under "poultry" and found what I was looking for: *Chickens—$2.50 each. You catch them. Call R. L. Holcombe, Stony Fork.* The price was right and Stony Fork was only a couple of miles away. I hopped in my truck and drove over the mountain. There were a half-dozen chickens running loose around Mr. Holcomb's place, including some small mousy brown hens and a couple of magnificent roosters. Mr. Holcomb told me they were gamecocks, and that he had once been a serious cockfighter but now he only kept a few birds for pets. After chasing them around his yard for an hour, I managed to catch a rooster and a couple of hens, but more importantly, I got him talking about chicken fighting. That's when I became aware that I was living in the midst of a world that was largely invisible to my fellow college professors.

Mr. Holcombe told me that if I wanted to learn about game chickens, I needed to talk to Fabe Webb, a legend among western North Carolina cockfighters. I gave him a call, and, to my surprise, he invited me over. A big man in his seventies, with red hair and a ruddy face, Fabe lived half a mile up a rough dirt drive in a small white house surrounded on three sides by the Pisgah National Forest. You could not see his place from the road, but you could hear the birds a long way off. He had dozens of gamecocks, resplendent animals—some deep maroon, almost purple, with iridescent green hackles; some pure white; others black with orange necks—all individually housed in wire cages.

Fabe loved to talk chickens, and I began to drop by regularly. He would take me through his chicken yard, explaining the bloodline of each rooster, tossing a bit of grain here and there, clucking to his birds. Every now and then, he would point out a battle-scarred old gamecock, perhaps with only one good eye, and Fabe would puff up a bit, like one of his roosters, and say, "Now that one's a six-time winner." The truth was that Fabe had given up cockfighting years before I met him, though he still went to an occasional derby, sat in the bleachers, and bet on a couple of matches.

I was completely baffled. How could a person who loved birds so much participate in a brutal and illegal bloodsport that always ends in death? I brought up this apparent contradiction one afternoon while were sitting in his kitchen sipping some of western North Carolina's finest homemade corn whisky. Fabe offered to take me to a fight. He said he wanted to show me that cockfights were not the bloodbaths that city boys like me thought they were. It never happened. I kept putting him off, until one day, I read in the paper that Fabe had died. I think it was a heart attack.

Shortly after Fabe's death, I was called into the office of the dean of the college that had hired me as a temporary instructor. He offered me a permanent job, but said that if I wanted the position, I would need to finish my PhD in psychology. All of a sudden, watching the chickens that I bought from Mr. Holcombe went from a casual pastime to my ticket to the tenure track. After reading up on poultry behavior, I convinced my advisor that I should write my dissertation on chickens. I was particularly interested in breed differences in the behavior of baby chicks, including gamefowl chicks. Two months later, I was running a minihatchery in my basement. I did not have any trouble getting fertile Rhode Island Red and White Leghorn eggs from the university's poultry science department, but getting my hands on purebred gamecock eggs was another matter. After contacting friends of friends of friends, I located a couple of Tennessee rooster fighters who were happy to oblige. And one of them, a man named Jim, invited me to tag along to an upcoming derby over in North Carolina. This time, I went.

FIVE-COCK DERBY IN MADISON COUNTY

The fight was at the pit near the abandoned Ebbs Chapel School in Madison County; from the outside, it looked like a big barn. We paid the admission fee, and Jim went over to talk to some of his buddies while I tried to make myself invisible. The air was hazy with cigarette smoke and heavy with the smell of coffee and hamburger patties frying on the grill at the refreshment stand. A hundred and fifty or so people were sitting in the bleachers or milling around. Among the spectators was a man in a wheelchair sitting next to his wife and son, who looked to be about twelve years old. Above the general din of conversation, the voice of the pit owner boomed over the PA, instructing the handlers awaiting the next match to bring their roosters to the main pit.

Each of the handlers was carrying an odd-looking rooster. They looked strange because their combs and wattles had been removed and the feathers on their backs and sides trimmed so the birds would not overheat during the fight. One rooster was deep red. His opponent was black with pale yellow hackles. Like high school wrestlers, the cocks were matched for weight. Gaffs, the curved pointed steel blades that make cockfights deadly, were attached to the stubs of the natural spurs on the legs of the chickens with leather and waxed string. As was customary back then in Appalachia, they were fighting long heels—two-and-a-half-inch gaffs—rather than the shorter gaffs favored in the North or the slasher blades worn only on the left foot that Filipino and Hispanic cockfighters prefer.

A bald guy in the bleachers yelled out to no one in particular, "I'll lay a 25-to-20 on the Grey." A younger man sitting across the pit pointed to him and said, "You're on." Jim nodded toward a handful of men hanging quietly together in a corner. He told me they were the high rollers who made big bets among themselves.

The referee was a fifty-year-old black man they called Doc, who by day was a school janitor. He gave the signal: "Bill em up." The handlers cradled their birds in their arms and brought them together in the center of the pit, a round arena about fifteen feet in diameter surrounded by a three-foot wire fence. Once the cocks saw each other up close, the adrenaline kicked in, their hackles flared, and they went for one anothers' eyes. After the birds pecked at each other's heads for a few seconds, the handlers separated them and retreated behind the long-score lines drawn in the dirt. The handlers squatted down, holding their birds, waiting. Doc yelled, "Pit 'em!" The handlers released the roosters and the fight began. All I saw was a blur of feathers.

Ten minutes later, the red cock's handler tossed his rooster's limp body into a barrel of dead chickens. By the time bets were paid off, two more roosters were readied for the next fight, as I heard Doc say, "Pit 'em!"

I dragged myself home at three in the morning and tossed and turned all night. I was trying to figure out what it all meant. Bob Dylan has a line that goes, "Something is happening here but you don't know what it is—do you, Mr. Jones." That's how I felt.

The next morning at breakfast, I told Mary Jean that I was going to shift gears and become an ethnographer for a while. I was certainly not the first researcher who had tried to make sense of rooster fighting. Most of the others have been anthropologists looking for underlying meaning—totems, myth, and symbolism. In 1942, Gregory Bateson and Margaret Mead wrote of cockfighting in Bali, "The evidence of regarding the fighting cock as a genital symbol comes from postures of men holding cocks, the sex slang and sex jingles, and from Balinese carvings of men with fighting cocks." The British anthrozoologist Gary Marvin interpreted the Spanish cockfight as a celebration of manhood, and Princeton sociologist Clifford Geertz argued that the function of rooster fights in Bali is to confirm the status hierarchy among the men in rural villages. More recently, in an essay titled "Gallus as Phallus," University of California anthropologist Alan Dundes argued that a cockfight is actually a "homoerotic male battle with masturbatory nuances."

While I found these quasi-Freudian notions interesting, they provided no insight into the question I was interested in—how could such apparently normal people be involved in an activity that most Americans, myself included, viewed as an exercise in sadism? But to figure out the paradox of why these seemingly good people do seemingly bad things, I had to learn about their sport. I had to step back and assume the mantle that neurologist Oliver Sacks called an "anthropologist from Mars." Over the next two years, I racked up a couple thousand miles driving the back roads of east Tennessee and western North Carolina, interviewing cockfighters, photographing their kids and roosters—usually together—and gathering data, checklist in hand, at clandestine rooster fights. Along the way, I learned a lot about how people think—and avoid thinking—about the way we treat animals.

THE CULTURE OF COCKFIGHTING: A PRIMER

Pitting roosters against each other is one of the oldest and most wide-spread of traditional sports. The chicken as we know it was domesticated from several species of wild Asian jungle fowl about 8,000 years ago. Wild jungle fowl are inveterate fighters; humans were probably staging bouts be-tween roosters from the time they began keeping chickens around for meat and eggs. Cockfighting originated in Southeast Asia and soon diffused to China, the Pacific Islands, the Middle East, and eventually to ancient Greece and Rome, where young men were required to attend cockfights to learn the meaning of courage. The sport took hold in Europe, becom-ing especially popular in Spain, France, and the British Isles. Columbus brought chickens—and probably cockfighting—to the New World, where it spread rapidly through North and South America.

To understand cockfighting, you must start with the chickens. Cock-fighters are obsessed with bloodlines. They talk endlessly about the merits of crossbreeding, linebreeding, and inbreeding. They can tell you about F1 and F2 generations—just like your high school biology teacher. I visited breeders who would pull out notebooks that went back decades. They knew who sired whom and which hens produced good shufflers and cut-ters. Now they keep these records on computerized databases.

There are hundreds of strains of roosters. They have great names: Blue Faced Hatches, Kelsos, Arkansas Travelers, Allen Roundheads, Madigan Grays, Butchers, Clarets. When breeding their battle cocks, rooster fight-ers are looking for the perfect combination of three traits. The first is cutting—the ability of a cock to deliver accurate strikes to its opponent's body, to puncture the lungs or heart. The second is the ability to put power behind the blows. But by far the most important trait, the one that gets breeders misty-eyed, is what they call true grit, or more commonly, gameness. I ask Johnny, a third-generation cocker, to tell me how I could explain gameness to my animal rights pals. "Gameness," he said, "is their heart. Their desire to fight to the death. Your barnyard rooster is cow-ardly. It can't take the steel off the gaff. Gameness is the drive to beat the opponent. It is so instilled in the true game rooster that he is going to give everything he has, to his last breath."

While cockers are as concerned with bloodlines as the members of the Westminster Kennel Club, all of them will tell you that good genes are not enough to make a great fighting cock. The roosters must also be raised right. Johnny's stance on the poultry version of the nature-nurture debate is that a roosters' fighting ability boils down to 85% genes and 15% conditioning. Several weeks before a derby, cockfighters put their roosters on a prefight diet called the *keep*. Every cockfighter has his own secret system. Johnny would give his roosters vitamin supplements during the keep and allow them to spend a few hours every day foraging for bugs on fresh grass. Some cockers start adding a pinch of strychnine to their rooster's food during the keep; they think it thickens a rooster's blood. Others dose their roosters with antibiotics, testosterone, or stimulants. (Johnny tried Dexedrine a couple of times but he gave it up when he found the drug made his roosters act crazy in the pit.)

Cockers think of their birds as athletes, and they have developed physical conditioning regimes to develop stamina and quickness. Johnny would work his birds in the morning and then again in the afternoon. He had a padded exercise bench where he would put them on their backs so they would learn to right themselves quickly, and he would practice "flirts" by flipping them backward to develop their wing and back muscles. Like iron pumpers at a Gold's Gym, Johnny's roosters had an exercise for each muscle group, and he kept track of each rooster's daily reps as he put it through its paces. In the weeks before a big derby, he would spend six hours a day conditioning his animals.

EVEN COCKFIGHTS HAVE RULES

Every sport has rules, and each of the world's cockfighting cultures has its own traditions that govern how fights are conducted. In Andalusia, for example, metal gaffs are not attached to the roosters' spurs, so Spanish cockfights are usually not lethal. The most common type of cockfight among cockers in the American South is called a *derby*. In a derby, each cockfighter enters a preset number of roosters in series of round-robin matches. The basic set of guidelines, referred to as Wortham's Rules, have been in place since the 1920s and are still used in cockpits from the hills of Kentucky to the plains of west Texas.

The rules are complicated—you don't just put the roosters together and let them have at it. During fights, there are two handlers, two chickens, and a referee in the pit. The referee controls the action. The handlers, who often are not the actual owners of the birds, bring the roosters close enough to peck at each other's faces. Then they place the birds on the ground eight feet apart, facing each other. On the referee's command, the roosters are released. Instantly and silently, they charge toward each other in a flurry that Clifford Geertz described as "a wing-beating, head-thrusting, leg-kicking explosion of animal fury so pure, so absolute, and in its own way so beautiful, as to be almost abstract, a Platonic concept of hate."

Within twenty or thirty seconds, the gaffs usually become entangled in the body of one or both roosters, and they collapse to the ground in heap. The referee signals the handlers to step forward and disentangle their birds; the handlers have twenty seconds to get their birds ready for the next pitting. A good handler knows the anatomy of injuries and, like the corner man in a prize fight, can bring a hurt animal back for the next round. He may put a little water on the cock's face, or blow on its head to settle him down, or just put him on the ground and leave him alone so the bird can sling a blood clot out of his throat. At the end of the rest period, the fight resumes. And the process repeats until there is a winner.

Sometimes a rooster will win a cockfight by killing its opponent outright or by beating it up so badly the opposing handler throws in the towel. But more often, the victor is determined by a complicated set of rules called the count. Above all, cockers value a gamecock's drive to fight no matter if its lungs are punctured, its spine shattered, or its vision growing dim. The count system helps ensure that the gamer rooster, the one that keeps fighting when all seems lost, wins. When a rooster stops attacking because he is injured or exhausted, the opposing handler says to the referee, "Count me." If the other rooster does not attack for four successive pittings, the rooster whose handler "has the count" is declared the winner. But if the bloodied and battered rooster makes the slightest effort at aggression, even a weak peck toward its adversary, the count starts over. A typical gaff fight lasts about ten minutes, but sometimes a rooster will get in a lucky blow and hit a vital organ and it is over in seconds. On the other hand, a single fight can go on for an hour or more. To keep the spectators from getting bored, long fights are moved to a secondary or "drag pit" which frees up the main arena for a pair of fresh birds.

In recent years, the influx of immigrants has affected cockfighting just as it has many other aspects of American culture. The weapon of choice of the new breed of cockers is the knife. These artificial spurs, unlike ice pick–shaped gaffs, have razor-sharp edges. Filipinos like the long knife, which looks like something you would use to dismantle a porterhouse at Ruth's Chris Steakhouse. Mexicans prefer the short knife, which has a one-inch blade honed on both sides. Old-timers aren't keen on these new developments. They claim the knife makes it easy for a less courageous rooster to win a fight by way of a lucky blow.

Cockfights are not nearly as bloody as you would imagine. Eddy's wife, a food service manager, describes the first time he took her to a fight at the Del Rio pit. "I was surprised. It was not the blood and guts I thought it would be. And the restrooms were clean. So was the kitchen and the food was good. They made everything from scratch."

She is right about the gore. The wounds inflicted by the steel gaffs do not bleed much on the outside, and the cock's feathers conceal most of the seepage. In addition, a game rooster's blood is more efficient at clotting than the blood of other types of chickens. In the parlance of cockfighters, game roosters can "take a lot of steel." Cockers have a lexicon of injury. A rooster that has had its lungs punctured emits a creepy rasping sound called the "rattles." When a cock flops around with spinal cord damage, it is said to be "uncoupled." I once took the carcass of a rooster that had been killed in a fight to the state veterinary pathology lab for autopsy. The pathologist, it turned out, had been involved in cockfighting when he was growing up in Oklahoma. When he cut the cock open, he counted nineteen holes in its body. The fatal blow was to the throat.

Fights are usually straightforward affairs; almost all the losers die and almost all the winners live. But every now and then, the unexpected happens. A rooster may refuse to fight and just walk around aimlessly with a bewildered look in his eye. Occasionally a cock will humiliate its owner by turning tail and running around the pit, squawking. Then there are the man-fighters, cocks that spin around and go for their handler instead of the other rooster. One night I saw a Grey nail a handler in the thigh with both gaffs, two-and-a-half inches of ice pick. The man turned ashen and collapsed.

Notch one up for the chicken.

One of the biggest surprises of my foray into the clandestine world of rooster fighters was how open it was. Most Appalachian cockfighters made no real effort to conceal their involvement in their sport, even though it was against the law. As I drove around the region, I would see fields containing hundreds of gamecocks, some tethered to barrels, others in little wooden houses that looked like Boy Scout pup tents, all of them in full view of rural roads or even interstate highways. How did they get away with it?

Simple. In the 1970s, being a cockfighter in Appalachia was about as illegal as being a litterbug. True, cockfighting was against the law in both North Carolina and Tennessee, but it was a misdemeanor, and local sheriffs generally turned a blind eye. On the rare occasions when raids occurred, the participants were invariably slapped on the wrist and given a $50 fine. No one ever went to jail. Sure, some sheriffs were on the take, but others figured it was better to have the fights in established pits where it was in everyone's interest to be on good behavior. These pits had implicit codes of conduct to avoid altercations: no drinking or drugs around the pit, don't welsh on bets, the referee's word goes. Pit owners also did their best to avoid trouble with their neighbors. The owner of the Ebbs Chapel pit would take donations at the Saturday night fights for the Baptist church up the road.

In short, the local law figured it was preferable to let the pits operate than force the sport more deeply underground. They may have been right. Cockfights involve a potentially explosive combination of testosterone and money. I was afraid only once at a cockfight, and it was at a ragtag event called a brush fight. Brush fights are informal affairs held in barns or in the piney woods, arranged by a few last-minute phone calls, with no entry fee, no paid referees, and no rules against alcohol. The fight was in west Knoxville on a hot afternoon and the beer was flowing. The referee was a spectator who was drafted from the crowd and did not really know what he was doing. Two handlers, who had both been drinking, squabbled over one of the ref's calls. In a regular pit, they would have been ejected immediately. Not so in a brush fight. The argument escalated until one guy grabbed a beer bottle, broke off the neck, and went for the other handler, catching him in the shoulder. With blood dripping down his arm, the man who had been stabbed grabbed his chicken and stomped off. When I heard him slur, "I'll teach that asshole a lesson. I've got a shotgun in my truck," I started looking for an escape route. Unfortunately, I had gotten to the fight early and was completely hemmed in by a dozen pickup trucks. However, everyone else was as averse to drunks with shotguns as I was. Within minutes every vehicle cleared out, and I was headed home, sweating, heart beating and hands shaking, thinking that maybe I was not cut out to be an anthropologist after all.

JUSTIFYING THE UNJUSTIFIABLE

Most people imagine that cockfighters are low-life scum who peddle crystal meth when they are not gleefully torturing animals. I found, however, that the most psychologically interesting thing about rooster fighters is how boringly normal they are. Nearly all the cockers I have known led—aside from their devotion to a brutal bloodsport—ordinary lives complete with mortgages, wives, children, and day jobs.

Suzie, a Louisiana animal protectionist I know, worked tirelessly for years to ban cockfighting in her state. Her experience with rural Southern rooster fighters was similar to mine. She despises cockfighting. However, like Baptists who claim to hate the sin but not the sinner, she came to respect many of her opponents. She told me once, "Most of the cockfighters I met were God-fearing, polite, family people who were not out there shooting heroin or snorting coke. I am completely opposed to cockfighting, but that does not mean that cockfighters are evil people."

If cockfighters were sadistic perverts, it would be easy to explain their involvement in a cruel bloodsport. But given that most are not, how can they participate in an activity that is illegal and that nearly everyone in America thinks is immoral? The answer is that they construct a moral framework based on a mix of wishful thinking and logic in which cockfighting becomes completely acceptable. In this regard they are no different from any other person who exploits animals—hunters, circus animal trainers, even scientists and meat-eaters. There are several lines of argument that cockfighters use to justify an activity that most people find unjustifiable.

"The Most Humane Sport"

Most cockfighters deny that their sport is cruel. They tell you that that the fight itself is only a small part of their sport. They tell you it takes two years to raise a chicken from hatchling to battle cock and that the fight, which is often over in a few minutes, is just a fraction of what it means to be a cocker.

But what about the pain and suffering? My friend Johnny claims that the steel gaffs have taken the cruelty out of cockfighting. The gaffs, he argues, make it a fair fight because they equalize each rooster's chance of winning. If it were not for the gaffs, he says, the roosters would batter each other to death with their three-inch natural spurs. My neighbor Paul Ledford laid another argument on me one morning when I asked him over a cup of coffee how much pain was inflicted during a typical cockfight. He shook his head and said, "Chickens don't feel pain. Chickens are too dumb to feel pain."

Sometimes you also hear the opposite argument, that chickens are moral agents who choose to fight each other to the death. According to this thread of logic, it is cruel *not* to let roosters meet their destiny in the cockpit. In his book *Fighting Sports*, Captain L. Fitz-Barnard writes, "Where the agents are willing, there can be no cruelty. For the gamecock, the joy of battle is his greatest joy." Fitz-Barnard thinks that cockfighting is ethically superior to hunting or fishing because the whitetail buck or brown trout you kill does not have a choice in the matter. "Where you have unwilling participants there must be cruelty . . . no sane person can pretend that the fish enjoys being lured to death, often with live bait; that the fox or the hare likes to be hunted and torn to pieces; or that the birds and beasts prefer a lingering death from gunshot wounds."

I don't buy for a minute the argument that gamecocks choose to fight because they achieve the avian version of self-actualization in battle. No, they fight because their brains are hardwired by thousands of years of intense selection to plunge their spurs into other male roosters. Even if they wanted to flee, the close quarters of the pit make it impossible. Fitz-Barnard's comparison of cockfighting and hunting, however, does strike an uncomfortable chord. As many as 30% of the 120 million wild birds shot by hunters each year in the United States will fall from the sky wounded and fully conscious. While the lucky ones will be found and killed quickly, millions of others will die lingering deaths. Fitz-Barnard is right: There is much more suffering caused by the legal sport of recreational hunting than the outlawed sport of cockfighting.

"It's Only Natural"

A common variation on the "it's not cruel" justification for cockfighting is that roosters are instinctive fighters, just like lions are instinctive zebra killers. This is a variation on the naturalistic fallacy. Johnny laid it out for me: "What we do is an act of nature in a controlled situation. That rooster's going to fight if we are there or not. We make things as even as possible for them to perform an act of nature. We don't make these roosters fight. That's what they were put here for. It is their purpose." (This, by the way, is also the reason most cockfighters I met did not approve of dogfighting. As Eddy's wife told me, "Cockfighting is not like dogfighting. They have to make the dogs mean. But these chickens are born to fight. They do it regardless of whether you are there or not.")

I run across the naturalistic fallacy a lot. In explaining her opposition to animal research, an animal rights activist told me that AIDS was "nature's way" of reducing human overpopulation. A woman I met at a party recently said to me, "I just don't get vegetarians. Humans have been eating animals for millions of years. That's what cows and chickens were put here for." I did not point out that her justification for eating chickens was identical to Johnny's rationale for fighting them. I suspect she would not have seen the parallels.

"The Nicest People You Ever Saw"

Restrictions on cockfighting were originally based not on concern for animal suffering, but for keeping the riffraff down. The association between with cockfighting and other forms of criminal activity continues today. For example, the Humane Society of the United States links cockfighting with prostitution, identity theft, robbery, Mexican drug cartels, illegal gambling, bribery, gang activity, tax evasion, money laundering, immigration violations, hand grenades, and murder.

Cockfighters, of course, don't see it that way. They view themselves as a persecuted band of brothers held together by a common set of values that include hard work, competition, respect for cultural traditions, and love of chickens. They discount the charges of booze, dope, hookers, and money scams. Of animal protection organizations, Johnny told me, "They call us everything—pimps, drug dealers. To them we are the scum of the earth. They are smarter than we are. They know how to paint us to people who don't understand." OK, he admits, there are a few bad apples—the cheat who dabs poison on his rooster's hackles or illegally sharpens the edges of a long-heel gaff. But they are a small minority. As for 99% of cockers, Johnny says, "You will never find a nicer group of people. Where else can you go where that much money changes hands on a nod without any controversy? Cockers are gentlemen."

The Great Man Defense

Cockers supplement the "good people" defense with a rhetorical twist social psychologists call reflected glory. The logic goes like this: "If ____ was a cockfighter, it must be OK." The list of names they insert in the blank includes George Washington, Alexander Hamilton, John Adams, Alexander the Great, Woodrow Wilson, Andrew Jackson, Henry Clay, Hannibal, Caesar, Thomas Jefferson, Benjamin Franklin, and Abraham Lincoln, who was reportedly a referee. Genghis Khan and Helen Keller are sometimes added to the list, though I am skeptical about the latter. A long line of British royalty are claimed as fellow travelers by modern cockfighters who often refer to rooster fighting as the "sport of kings."

Cockfighting Builds Character

When Bobby Keener of Greensboro, North Carolina, was asked what kept him interested in cockfighting, he said, "It's the way that this animal would keep going until he has nothing else to give. How many people would do that? This bird will give you everything he's got until he's got no more and then he keeps giving it. It's what you call gameness or heart. That's what's kept me interested in it." I call this the moral model defense. To a cockfighter, a game rooster is the bravest creature on earth. That's why the gamecock is the mascot of the University of South Carolina football team. A cocker summed up the moral model argument when he wrote in *Grit and Steel*, "A gamecock is loyal to his family and himself—and he has the grit to back that loyalty. . . . It takes grit to be loyal—to your ideals, to your wife, to your husband, to your friends, to your country."

"I Love My Chickens"

For a chicken, a game rooster has it pretty good. Cocks are not usually fought until they are two years old, during which time they lead a life befitting a thoroughbred race horse. For the first eight or nine months, they have the run of the chicken yard. Once they hit puberty, the cocks have to be separated, but because breeders want their birds to get exercise, they are either tethered with a seven-foot "tie out" cord or kept in fairly large cages that allow them to move about. In addition to the organic corn that Johnny bought for them at the health food store, his roosters got hardboiled egg for breakfast and dined on fruit, salad greens, and pearl barley for lunch. Every other day, he would give them hamburger supplemented with cottage cheese. Johnny complains, "We give our roosters the best food, the best housing, the best hens. . . . And then they call us cruel."

Like every cockfighter I have ever talked to, Eddy Buckner is passionate about his roosters. He tells me he loves them, and I believe him. Like Fabe Webb, his eyes sparkle when he talks about his birds.

"But, Eddy," I say. "You claim to love these animals. You raise them practically by hand for two years, you spend hours every day handling them and exercising them. But then you take them to the pit on the weekend knowing full well that half of them are going to die before the end of the night, and then you just toss them in a barrel. I don't get it."

"You have to draw the line," he says.

"But don't you get attached to them?" I ask.

"Sure," he says.

"Do you ever name them?"

"Yes."

"Did you ever see a cockfighter cry over a dead rooster?"

"Never."

"I still don't get it," I tell him.

ANIMAL ACTIVISTS VERSUS ROOSTER
FIGHTERS: AN ASYMMETRY OF HATE

Cockfighters are completely convinced by these arguments—so convinced that rooster fighters have told me with a straight face they would like to bring animal rights activists to a derby so they could see what a great sport cockfighting really is. This idea is, of course, ludicrous. Every animal protectionist I know thinks that cockfighting is cruel, and nothing would ever change their minds. This leads to an asymmetry of animosity—animal activists hate cockfighters more than cockfighters hate animal protectionists.

Karen Davis, the founder of United Poultry Concerns, the country's only poultry rights organization, despises rooster fighting. She says that the sport is not about competition, but about male insecurity. The great irony of manhood, she tells me, is that men are scared of each other, that they are afraid that other men will see their touch of female sensibility. To her, cockfighting is a perverse adult version of the playground taunt, "My dad can beat up your dad."

When I ask how she feels about cockfighters. Karen says, "A chicken is about as tall as your knee, and they are at the total mercy of these men who tower over little birds and punish their bodies. And the ultimate punishment is putting them in a ring and making them beat each other up. Cockfighters force their violent, bloodlusting impulses on birds. And then they say they admire these animals and love them. I don't think there is a chicken in the world that would be grateful for that."

Is she right? Given a choice, would any chicken in the world volunteer to be a gamecock? Karen has forced me to revisit the question I asked myself while I was chasing the chicken trucks down I-40: Would I rather be a fighting rooster or a commercial broiler? To answer it, we need to look at the sorry lot of the industrial chicken.

WHAT WOULD YOU RATHER BE, A
GAMECOCK OR A BROILER?

The modern broiler chicken is a technological marvel. The chickens I followed down the interstate looked like Cobb 500s, one of the world's most popular types of meat chicken. The Cobb 500 was developed by Cobb-Vantress, a multinational corporation formed in 1986 as a joint venture between two corporate giants, Tyson Foods and Upjohn. Cobb-Vantress, which has operations in Europe, Asia, South America, and Africa, also produces the Cobb 700, designed for a high proportion of breast meat; the Cobb Sasso 150, aimed at the free-range and organic market; and the Cobb Avian 48, which is advertised as having "high livability" and particularly suitable for "live bird markets seen in some parts of the world."

These birds are meat machines. The average Cobb 500 breeder hen will produce 132 chicks by the time she is "depleted" at the age of fifteen months. The chicks' lives will be much shorter than their mother's. In 1925, it took 120 days and ten pounds of feed to produce a scrawny two-and-a-half-pound bird. Now chickens are slaughtered when they are six or seven weeks old, at which time they will weigh nearly five pounds. While a Cobb 500 grows nearly five times as fast as your grandmother's barnyard chicken, it eats less food. Cobb-Vantress projects that soon it will only take a pound and a half of feed to produce a pound of chicken. Even better from an industry perspective, there is less waste on a modern chicken. Once its feathers are plucked, its feet and head chopped off, its guts scraped out and blood drained, 73% of a Cobb 500's carcass will be "eviscerated yield."

Cheap meat comes at a cost. A broiler chicken's bones cannot keep up with the explosive growth of its body. Unnaturally large breasts torque a chicken's legs, causing lameness, ruptured tendons, and twisted leg syndrome. According to Donald Broome, professor of animal welfare at Cambridge University, severe leg pain in chickens is the world's largest animal welfare problem. Arthritis, heart disease, sudden death syndrome, and a host of metabolic disorders are prevalent among industrial broilers.

The living conditions of the animals destined to become chicken nuggets are Dante-esque. The chicks will never see sun nor sky. Because they are so top-heavy, broiler chickens spend most of their day lying down, often in litter contaminated with excrement. As a result, many will develop breast blisters, hock burns, and sores on their feet. A chicken "grow-out house" can be 600 feet long and 60 feet wide and hold 30,000 birds. The broiler houses are humid, the air laced with ammonia produced by the action of microbes on the accumulated urine and the excrement of tens of thousands of birds. The gas burns the lungs, inflames the eyes, and causes chronic respiratory disease.

When the broilers reach five or six pounds, it is time for their trip to the processing plant. But first you have to catch them. Crews of low-paid chicken catchers invade the sheds at night. Wearing masks and throw-away coveralls, they grab the chickens by the feet, five birds in each hand, and stuff them into holding drawers. It can take all night for a team of catchers to load 30,000 chickens into wire crates, during which time each catcher will lift fifteen tons total of flapping, squawking bird. It is nasty work for both people and animals. The catchers get scratched, pecked, and covered with shit. As many as 25% of the birds will be injured in the catching and loading process. My colleague Bruce Henderson, now a developmental psychologist, signed onto a chicken-catching crew when he was in high school. He lasted six days.

Isn't there a better way to snatch up chickens? Enter the mechanical chicken harvester. There are several varieties of these behemoths including one with giant rubber fingers. One of the more popular is the PH2000 made by the Lewis/Mola Company of Bennettsville, South Carolina. The machine is forty-two feet long, weighs 16,000 pounds, and can clean 24,000 birds out of a broiler house in three and a half hours. It works more or less like a two-headed dustpan. The machine gingerly sweeps two metal ramps called catch heads through the flock of chickens. They nudge each other on to ramp where they are moved toward a conveyor belt, which transports them directly into wire transport cages for the trip to the processing plant. The Humane Society of the United States says mechanical harvesters are easier on the birds than hand catching. Chickens hate being grabbed by humans and held upside down by their sore legs. They do not seem as stressed as they march clueless into the maw of the PH2000 and are excreted moments later, a bit dazed, into wire transport cages. The company reports that mechanical harvesting results in 60% fewer broken wings and a 99% reduction in broken legs. Despite these presumed advantages, the vast majority of the broiler chickens consumed each year in the United States are still captured by hand.

Once the transport cages are loaded, the trucks hit the road and take our Cobb 500s to the processing plant, where they will be dumped from their crates, shackled tightly by their legs with metal cuffs, and hung head-down from a conveyer belt. Then, upside-down and wings flapping, their heads will be passed through an electrified water bath. For seven to ten seconds, electricity will travel through their bodies, hopefully stunning the animals. The next stop is the neck-cutting machine where sets of ro-tating blades sever the chickens' carotid arteries. Once the chickens bleed out, their bodies will be dunked in the scald tank. An efficient production system with two evisceration lines can process 140 birds per minute. The system, however, does not always work. Some birds are not completely stunned before their throats are slit, and if the neck-cutting blades miss a chicken's carotid arteries, it will be conscious when it is submerged in the hot water of the scald tank.

There you have it. Your average east Tennessee gamecock chick will be pampered during its two-year life. For the first six months, it will run free. Then it will have a lawn to loll about and a private bedroom to sleep in. The rooster will get plenty of exercise, eat better than some people, and have a chance to chase the hens around. The downside is that one Saturday night, he will feel the sharp pain of the Mexican short knife slic-ing his pectoral muscles, or perhaps get a long heel gaff in the throat; he will die in the dirt after a fight that lasts anywhere from a few seconds to over an hour, during which men sporting baseball caps shout out odds to each other. His chances of seeing the sun rise Sunday morning are 50:50.

In contrast, our Cobb 500 chick will live in unimaginable squalor, legs aching, lungs burning, never glimpsing the sky or walking on grass or having sex or pecking at bugs, eating the same dreary poultry chow day after day for its forty-two-day life, when it will be jammed into a crate on an open truck and carted to the plant where it will be suspended upside down, electrocuted, and its throat slit. Its chances of seeing the sun rise are zero.

Karen Davis tells me that no chicken in the world would want to live the life of a fighting rooster. I'll lay 25-to-20 that she is wrong.

HOW MONEY AND SOCIAL CLASS AFFECT OUR PERCEPTIONS OF CRUELTY

Looked at objectively, it is hard to deny that there is less suffering caused by cockfighting than by our apparently insatiable demand for chicken flesh. It is likely that 10,000 or 20,000 chickens have their necks slashed in a mechanized processing plant for each gamecock that dies in a derby. And there is the inconvenient fact that the life of a fighting cock is fifteen times longer and infinitely more pleasurable than the life of a broiler chicken. Why then is it legal for us to kill 9 billion broiler chickens every year, but cockfighting can get you hard time in the federal penitentiary?

It is, in part, a matter of money and power. The National Chicken Council is the trade association of the poultry industry. Its members include the corporations that produce 95% of the the broilers consumed in the United States, and the organization works tirelessly to keep the government at arm's length. As a result, factory-farmed chickens are exempt from virtually all federal animal welfare statutes including the Humane Methods of Slaughter Act, enacted by Congress in 1958 to ensure that animals destined for the table will not unduly suffer as they are being killed.

While the National Chicken Council furthers the interests of the broiler industry, organizations such as United Poultry Concerns, PETA, the Farm Sanctuary, and the Humane Society of the United States

(HSUS) are the voice of America's chickens. Of these, the HSUS has the most political clout. With an annual revenue of $100 million dollars, and assets valued at $190 million dollars, the HSUS is the 800-pound gorilla of the animal protection movement.

In 1998, the HSUS decided to take on the cockfighters. In response to political pressure from animal protection groups, the last states caved. Louisiana was the final holdout. As the state sought to improve its image in the wake of Hurricane Katrina, the influence of the cockfighting lobby in the legislature waned, and in 2007 Louisiana Governor Kathleen Blanco signed a bill that closed the loophole that excluded chickens from the state's animal cruelty laws. As of August 15, 2008, cockfighting became illegal in every state. State cockfighting laws, however, are inconsistent. In Florida, cockfighting will get you five years in jail and a $5,000 fine, but in neighboring Alabama it is a $50 misdemeanor. Now that cockfighting is banned everywhere, HSUS's mission is to rally their members to pressure lawmakers to make cockfighting a felony in every state.

The war on cockfighting is about cruelty, but the subtext is social class. The eighteenth-century movement against blood sports was directed toward activities that appealed to the proletariat, such as bull-baiting and cockfighting, rather than the cruel leisure pursuits of the landed gentry, such as fox-hunting. It's no different today. Cockfighters come from easy groups to pick on—Hispanics and rural, working-class whites. Animal activists, on the other hand, tend to be urban, middle-class, and well educated. They dismiss rooster fighters as a motley group of shit-kickers and illegal aliens.

Duke University's Kathy Rudy, an avid animal advocate and dog rescuer, is troubled by the divide between animal activists and the working class. In an essay that appeared in the *Atlanta Journal Constitution* shortly after Atlanta Falcons quarterback Michael Vick pled guilty to charges of dogfighting, Kathy pointed out that our society is much more likely to criminalize forms of animal abuse that involve minorities and the poor than animal cruelties that affect the wealthy. The comedian Chris Rock made the same point on national television in response to a photograph of Sarah Palin, then governor of Alaska and an avid big-game hunter. He told

Letterman, "She's holding a dead, bloody moose. And Michael Vick's like, 'Why am I in jail?' They let a white lady shoot a moose, but a black man wants to kill a dog? Now that's a crime."

You see the same thing in thoroughbred racing. According to an investigation by Jeffery McMurry of the Associated Press, more than three horses per day died at racetracks in the United Sates in 2007—over 5,000 horses between 2003 and 2008. Yet, a Gallup poll taken after the death of Eight Belles, the horse who collapsed during the 2008 Kentucky Derby and had to be immediately euthanized, found that most Americans oppose any ban on thoroughbred racing. Like cockfighting, horse racing represents a confluence of gambling and suffering. But unlike cockfighting, thoroughbreds are the passtime of the rich.

COCKFIGHTING AND HUMAN MORALITY

I am conflicted over many moral issues involving animals. Cockfighting, however, is not one of them. I liked many of the cockfighters I met in the course of my research, but, as was the case with slavery, their sport is a cruel and unjustifiable anachronism. It is time for rooster fighters to close down the pits and swap their gaffs for golf clubs and bass boats.

I am, however, still disturbed by what our attitudes about cockfighting say about moral hypocrisy and the fallibility of common sense in our relationships with other species. While the great chicken-eating public (of which I am a member) will sleep easy tonight knowing that cockfighting is now banned in every state, teams of chicken catchers from Maryland to California will enter darkened broiler houses and stuff 35 million terrified birds into wire crates in preparation for their journey to the processing plant tomorrow morning.

Back when I was doing my doctoral research on cockfighting, I took an organizer from Amnesty International named Tony Dunbar to a derby at the Ebbs Chapel pit. Tony's day job was trying to save convicted murderers from execution. As we were driving home at 2 AM reeking of tobacco smoke and greasy cheeseburgers, I asked him what he thought of our evening rubbing shoulders with some of western North Carolina's ace rooster fighters. After a pause, he said, "To me, cockfighting is a small moral problem."

A lot of people would disagree. But I find it hard to escape the conclusion that, in comparison to the suffering that goes into producing a six-piece Chicken McNugget Happy Meal, he was right.

7

Delicious, Dangerous, Disgusting, and Dead

THE HUMAN-MEAT RELATIONSHIP

> You ask me why I refuse to eat flesh. I, for my part, am astonished that you
> can put in your mouth the corpse of a dead animal, astonished that you
> do not find it nasty to chew hacked flesh and swallow the juice of death
> wounds. —J. M. COETZEE

> All normal people love meat. You don't win friends with salad.
> —HOMER SIMPSON

Staci Giani is forty-one but looks ten years younger. Raised in the Connecticut suburbs, she lives with her partner, Gregory, in a self-sustaining
eco-community deep in the mountains, twenty minutes north of Old Fort,
North Carolina. Staci radiates strength, and when she talks about food,
she gets excited and seems to glow. She is Italian-American, attractive,
and you want to smile when you talk to her. She tells me that she and

Gregory built their own house, even cutting the timber and milling the logs. I think to myself, "This woman could kick my ass."

Staci wasn't always so fit. In her early thirties, Staci's health started going downhill. After twelve years of vegetarianism, she began to suffer from anemia and chronic fatigue syndrome, and she experienced stomach pains for two hours after every meal. "I was completely debilitated," she tells me. "Then I changed the way I ate."

"Tell me about your diet now. What did you have for breakfast today?" I ask.

"A half-pint of raw beef liver," she says.

Animal activists sometimes claim that Americans *en masse* are forsaking barbecued ribs and buffalo wings for garbanzo bean burgers and tofu nut loaf. It is true that an increasing number of people believe that animals are entitled to basic rights, including, one presumes, the right not to be killed because you happen to be made out of meat. But, despite our stated love for animals, Americans eat 72 billion pounds of animal flesh each year, and only a tiny fraction of people in the United States are true vegetarians. We kill 200 food animals for every animal used in a scientific experiment, 2,000 for each unwanted dog euthanized in an animal shelter, and 40,000 for every baby harp seal bludgeoned to death on a Canadian ice floe. And, in spite of what you sometimes hear, over the past thirty years, the animal rights movement has not made much of a dent in our desire to dine on other species.

In most cultures, meat is a symbol of wealth, and as a country gets richer, its citizens want to eat more of it. Since the 1960s, per capita consumption of meat has increased sixfold in Japan and fifteenfold in China. Frank Bruni, former *New York Times* restaurant critic, captured the luxury of meat when he described a $90 steak he had for dinner one evening at a Manhattan chop house. It was, he wrote, "a sublime hunk of glorious meat that you dream about for hours later, pine for the next day and extol in a manner so rapturous and nonstop that friends begin to worry less about your cholesterol than your sanity."

My revelation about the transcendent pleasures of meat occurred when Mary Jean and I were treated to an obscenely expensive dinner by a couple of old friends who were celebrating a milestone in their lives. Two waiters for our table; five wines, selected by the sommelier to match the food; a teaspoon of lemon shaved ice to perk up the taste buds between the soup and the fish course. The entrees were up to the chef, but there were a couple of appetizer options. Mary Jean chose duck confit. I went for the pork belly.

I had never tasted pork belly, but I remembered that the local country music station used to announce their going price during the noon farm and home show. The piece of pork belly on the plate in front of me was a no-frills chunk of braised fat. One bite and my ideas about meat changed. I once stood for ten minutes in a museum staring at a painting by Mark Rothko, trying to figure out why anyone would think that an all-black canvas was art, but then something clicked and I suddenly got it. I had the same response to the taste of that pork belly. The Rothko and the pork belly had the same Platonic purity. One was distilled blackness, the other the essence of meat.

What is it about meat that brings out the contradictions and conflicts in our thinking about other species? The problem with meat is that while it tastes good, it can be unhealthy, it is disgusting, and it involves killing animals.

WHY IS MEAT SO TASTY?

When asked in a Gallup poll what they would choose for their "perfect meal," twenty times more people picked a meat entrée than a main dish made from plant foods. Even half of vegetarians admit that they sometimes have meat cravings. Why are humans so drawn to the taste of flesh? The reason is clear—we evolved from a long line of meat-eaters.

Chimpanzees, our closest living relatives, love the taste of meat. Richard Wrangham, a Harvard University primatologist who has studied chimps for decades, says he has never heard of a wild chimpanzee who was not crazy about meat. Female chimpanzees will readily exchange meat for sex, and Craig Stanford, professor of anthropology at the University of Southern California, has seen juvenile chimpanzees pathetically staring up at meat-eating adults sitting in a tree, desperately hoping that a few drops of blood will fall their way. Adult chimps hunt rats, squirrels, small antelope, baboons, and even baby chimpanzees, but their favorite meal is red colobus monkey. Chimpanzee carnivory is brutal. The chimps of Taï National Park in the Côte d'Ivoire kill by disembowelment, while the Gombe chimpanzees of Tanzania are prone to torture their victims, tearing their limbs off or smashing their heads against tree trunks or rocks. In the Kibale forest of Uganda, chimps usually begin their meal by eating the viscera and organs of their prey while it is still alive.

Oddly, while chimps enjoy flesh, they don't eat all that much of it. Meat makes up only 3 or 4% of a typical chimpanzee's diet, and even the most voracious chimpanzee meat-eaters only average a couple of ounces a day. By 2.5 million years ago, our hominid ancestors were eating more meat than do modern chimpanzees. This shift toward omnivory was accompanied by changes in body and brain. Compared to apes, modern humans have a relatively small gut, with less devoted to the colon and more to the small intestine. Our teeth also reflect a fleshier diet. Like chimpanzees and gorillas, the teeth of our early australopithecine forebears lacked the concave shearing blades of serious meat-eaters; their large flat molars were designed to masticate tough vegetation and grind up husks of seeds and nuts. Our teeth, in comparison, are a Swiss army knife of slicers, crushers, and biters.

The most important effect of omnivory was its influence on the evolution of the human brain. Many anthropologists believe that the shift to eating meat was a critical factor in the tripling of the size of our brains over a couple of million years. The idea that we evolved from meat-hungry apes is not new. Raymond Dart, who in 1924 discovered the first "ape-man" fossil, the Taung Child, in South Africa, wrote that our ancestors were bloodthirsty killers who delighted in "greedily devouring livid writhing flesh." Recent theories about the role of meat in human evolution are more sophisticated than Dart's, but the basic idea is the same. Craig Stan-

ford believes that it is not a coincidence that humans, chimpanzees, baboons, and capuchin monkeys—the primates that eat the most meat—are
also the most adept social manipulators. They are good at deception and
making alliances; they are skilled in the nuances of complicated interpersonal relationships. He believes that meat sharing jump-started brain
evolution by facilitating social intelligence.

Our forebears probably made several dietary shifts, first from a high-
fiber diet based on plants to meals that included more animal flesh. Then,
after the invention of agriculture, they went back to more vegetable-based
fare. The diets of hunter-gathering peoples show that humans can adapt
to an extraordinary range of foods, but they always included some meat.
Before snowmobiles and satellite TVs came to the Arctic Circle, 99% of
the caloric intake of the Nunamuit people of northern Alaska was derived
from animal products—foods like raw *muktuk* (whale skin and blubber),
fish, walrus, and fermented seal flipper. On the other hand, the !Kung
of the Kalahari Desert survived easily on a diet composed 85% of plants.
After surveying the diet of hundreds of hunter-gatherer groups, Loren
Cordain, a nutrition researcher at Colorado State University, found that
these groups obtained, on average, two-thirds of their calories from animal
flesh. Not a single hunter-gatherer society survived on a diet that was less
than 15% animal products.

Bobo Lee, whose real name is Robert E. Lee (seriously), agrees with
the theory that meat-eating is part of the natural order of things. BoBo
and his wife, Pam, run Po-Pigs BBQ, a small barbecue joint next to a gas
station forty-five minutes outside of Charleston, South Carolina. BoBo
was a commodities trader before he became a barbecue man. I stumbled
upon Po-Pigs while driving through what South Carolinians call the Low
Country. I only discovered later that road-food gurus Jane and Michael
Stern had named it one of the top five BBQ places in the United States.

In my experience, most barbecue restaurants disappoint. Not Po-Pigs.
The atmosphere is right—checkered plastic table cloths, a hand-printed
sign taped on the door saying "cash and checks only," and, most important,
no windows (always a good sign in a barbecue place). The pulled-pork is
sublime—moist, smoky, and sweet. While there are four squeeze bottles

of sauces on each table, they are distractions; the meat speaks for itself. When I complimented BoBo, he said to come back the next morning and we could talk meat. When I showed up at 9 AM, George Green, a retired army mess sergeant with a heavy Gullah accent, took me into the backroom and opened the top of the cooker. Through the smoke, I made out a couple dozen mahogany-colored bone-in pork butts that had been cooking all night at 200 degrees. They were ready to be taken off the heat. After letting them sit for a half-hour, BoBo and George don heavy insulated gloves and pull the hot meat apart by hand, mixing in a little homemade Pee Dee vinegar-pepper sauce. BoBo says that if you have to chop up the pork with a knife, you haven't cooked it slowly enough.

I asked BoBo why humans find meat so satisfying.

"That's how our ancestors survived." he said. "By killing animals. Hell, they even ate mastodon. It's been ingrained in our heads that sitting down and eating a good piece of meat is a sign of success. It makes your mind feel good; it makes your stomach feel good. There is nothing to me as satisfying as a warm, bloody, medium-rare steak."

But that does not explain why his wife, Pam, won't touch the stuff. She doesn't even like to look at raw meat. "Pam," I asked, "you have served a couple of hundred pounds of pork a day for the last ten years, but you don't eat meat?"

"Well, I eat fish."

She says, "I never liked the taste of meat. Even when I was a child, my mother would put it in my mouth, but I wouldn't swallow it. It was not so much the flavor; I didn't like the texture. Now they make a vegetarian hotdog so I am in hog heaven."

"How do you like the barbecue here?"

"I have never tasted it."

DEALING WITH THE DANGERS OF DEVOURING FLESH

Pam is unusual in her aversion to animal flesh. Most people are like her husband. Given the evolutionary history of our species, it makes sense that humans would naturally be attracted to meat. Biologically, you get a lot of nutritional bang for the buck. But the flip side is that meat is the most dangerous of the foods we eat. When ABC News surveyed Americans about the kinds of foods they were most afraid of getting sick from, 85% of people listed some form of meat, compared to 1% who mentioned a vegetable product.

Our fear of animal flesh is well founded. The problem is that we, too, are made of meat. And thus we are susceptible to the various bacteria, viruses, protozoans, amoebas, and parasitic worms that inhabit the creatures we eat. Fish carry at least fifty kinds of transmittable parasites. The flesh of cows, pigs, goats, and sheep potentially carry the bacteria *E. coli,* which causes 400,000 human deaths worldwide each year. Some researchers believe that the AIDS virus was first transmitted to our species through the practice of eating apes. Then there are insidious little prions—tiny pieces of protein that have the remarkable ability to reproduce in your cells even though they contain no genetic material. They are responsible for bovine spongiform encephalopathy (or "mad cow disease"), which slowly turns the brain into Swiss cheese. Prions also cause kuru, a neurological disease found among the Fore people of New Guinea that is transmitted by the consumption of the brain tissue of deceased relatives during funeral rites.

But, you ask, why don't lions and wolves get sick from eating raw meat? Harvard's Richard Wrangham chalks it up to cooking. He argues that the invention of cooking by *Homo erectus* 2 million years ago was the breakthrough that made the big brain possible by opening up a wider range of edible foods. In addition to adding flavor to their impala tenderloins, cooking also destroyed many of the pathogens that could make our ancestors ill. As a result, humans did not need to evolve the biological defense mechanisms that enable true carnivores to resist the toxins produced by bacteria that live in meat.

The addition of spices to the human diet also made meat less dangerous. Paul Sherman, a biologist at Cornell University, wondered why humans are unique among animals in spicing foods, particularly with inherently aversive substances like red-hot chilies, which can burn at both ends. Sherman and his students hypothesized that humans developed the taste for spicy foods because spices contain chemicals that retard the growth of harmful microbes. To test this idea, they analyzed thousands of traditional recipes from all over the world. In every country, meat dishes were spicier than vegetable dishes. Further, people living in hot and humid areas that are conducive to the growth of bacteria use more spices on their foods than people who live in cooler climates. In India, Indonesia, Malaysia, Nigeria, and Thailand, every meat dish is heavily spiced. (By the way, the most potent antibacterial spices are garlic, onion, allspice, and oregano.)

While cooking and spicing makes meat safer, eating flesh is still risky, particularly for pregnant women. Meat-borne pathogens such as *Toxoplasma gondii, Listeria monocyogenes, E. coli, Shilella dysenteriae,* and *Leptospira* can cause spontaneous abortion, stillbirth, and prematurity. But evolution has come up with defenses against foods that might harm a developing embryo—nausea and food aversions. Most women experience nausea and vomiting sometime during the first three months of pregnancy when their developing embryo is most susceptible to the harmful effects of toxins. Paul Sherman reasoned that because meats are the most dangerous of foods and fruits the safest, meat aversions should be more common during pregnancy than fruit aversions. His analysis of the food preferences of 12,000 pregnant women showed that this was true. Pregnant women are ten times more likely to develop aversions to meat than to fruits.

WHY SHEEP BRAINS ARE DELIGHTFUL IN BEIRUT AND REPULSIVE IN BOSTON

Pregnant women are not the only people with meat aversions. We all have them. My favorite treatise on meat is a slim volume published by a Dane named Peter Lund Simmons in 1859, *The Curiosities of Food or the Dainties and Delicacies of Different Nations Obtained from the Animal Kingdom.* Simmons chronicles the extraordinary range of animal flesh that humans eat, most of which I would not touch. He describes the joys of elephant's toes (pickle them in strong vinegar and cayenne pepper) and the gustatory delights offered by creatures such as porpoises, wombats, coffee rats, toads, bees, centipedes, spiders, sea slugs, and flamingos (the tongues of which are "extremely rich, much like that of the wild goat").

As an adventurous eater, I would rate myself a 7 on a 10-point scale. I have enjoyed the delights of sheep brain (which I ate regularly when I was a student in Beirut—fried is better than boiled), pig intestines, jellyfish, snapping turtle, sheep stomach stuffed with offal, grasshopper, thymus glands, black bear rump roast, alligator, and sun-dried iguana eggs. However, I find yogurt disgusting and sushi bland. I could not bring myself to eat cat, rat, bat, or chimpanzee. Neither would I eat *balut*—the Filipino delicacy that consists of a warm half-fledged duck embryo sipped from the shell. Nor could I gulp down, as did New York celebrity chef Anthony Bourdain, the beating heart of a cobra. But despite my failures as a gourmand, all of these foods are regarded as delicacies in some part of the world.

Why is the list of edible animals so long, yet, by comparison, the number of creatures whose flesh we regularly eat so short? One reason is availability. Jared Diamond points out in his book *Guns, Germs, and Steel*, that while most animals are edible, few are good prospects for large-scale agricultural production. For example, only fourteen of the world's 148 large terrestrial mammals have been domesticated. Your options in meat also depend on where you live. The meat counter at the supermarket where I shop only carries the standards—beef, pork, and chicken, with a few packages of lamb and a couple of kinds of fish thrown in. For the brave, there is liver. But, if you are reading this in Barcelona, you can trot over to la Boqueria, the cavernous central market on La Rambla. About halfway down the aisle on the right, you will find the viscera monger's stall. Get there early and it will be piled with shimmering internal organs—stomachs, brains, tongues, intestines, lungs, hearts, kidneys, even a couple of freshly skinned sheep heads.

Lack of availability, however, is only one reason why people avoid eating certain types of meat. Personal experience also comes into play. Like rats, humans have evolved a special ability to associate the taste of a food with nausea and vomiting. This was discovered by the psychologist Martin Seligman who developed an aversion to one of his favorite foods, steak with béarnaise sauce, after eating it on his birthday and then coming down with a virus and throwing up all night. Not surprisingly, learned aversions to meat are three times more common than aversions to vegetables and six times more common than fruit aversions.

The most important influence on whether we find a food delicious or disgusting, however, is culture. Daniel Fessler, an evolutionary anthropologist at UCLA, has studied food taboos across human societies. Because it is dangerous, Fessler reasoned that meat should be more frequently tabooed than plant-based foods. He and graduate student Carlos Navarrete collected information on forbidden foods in seventy-eight cultures. They found that perfectly edible meats were six times more likely to be forbidden than vegetables, fruits, or grains.

Why should it be easier to taboo meat than plant foods? Anthropologists love questions like this. As is often the case, there is a lot of speculation and little hard data. Some food anthropologists are functionalists who believe that taboos are adaptive. Pork, for example, is forbidden for both Muslims and Jews. Some of the functionalists believe that the taboo on pork serves to protect humans from trichinosis. Another functionalist view is that the prohibition on eating pigs was adaptive because swine competed with humans for the same types of foods. Similarly, the anthropologist Marvin Harris argues that the veneration of cattle among Hindus in India developed because cattle are more useful for plowing fields and producing milk and fuel (dried dung) than as a source of protein.

In recent years, functionalist explanations of meat prohibitions have not fared well. They do not, for instance, explain the geographic distribution of meat taboos. Why, for example, are cattle eaten in Pakistan where, as in India, they till soil and produce milk and fuel? Nor do they explain ecologically paradoxical taboos such as the prohibition against eating fish among desert dwellers such as the Navajo Indians of the American Southwest or the pastoral Masai in Africa. An alternative to the theory that meat taboos are adaptive is the idea that they result from the quirks of human psychology. I suspect that most meat taboos are simply the result of arbitrary cultural traditions and have no explanation other than the human tendency to copy one another.

If I am correct, under the right circumstances, our feelings about edibility of animals should sometimes change quickly, just as popular baby names do. This was indeed the case with a taboo against eating buffalo among the Tharu people of Nepal. The anthropologist Christian McDonaugh lived in a Tharu village between 1979 and 1981. During this time, McDonaugh regularly ate pigs, goat, fish, chicken, and even rats with the villagers—but never buffalo. Buffalo and other animals were slaughtered as part of religious rituals. But unlike the carcasses of chicken, pigs, and goats that were eaten after the rituals, the dead buffalo were dragged off and discarded. Twelve years later, McDonaugh returned to the village. He was shocked when he was offered a snack of buffalo meat toward the end of a long afternoon of beer drinking. The Tharu, it seems, had changed their attitudes. McDonaugh attributes the rapid decline of the buffalo taboo to several factors. First, the price of other meats had gone up, making buffalo a bargain. Second, the caste system was eroding. The population of the valley was becoming more diverse and the Tharu were exposed to people who did eat buffalo meat. Finally, the region was becoming more democratic and the Tharu were freer to express their political opinions and aspirations. For the first time, they felt they could eat whatever they wanted to.

DOGMEAT COOKIES, DOGMEAT STEW: A CASE STUDY OF A DIETARY TABOO

When a culture taboos a type of meat, even the idea of eating it becomes revolting. For most Americans, the idea of consuming dogmeat is particularly repulsive. The archaeological evidence, however, indicates that humans have been eating dogs for thousands of years. In many parts of the world, people have historically treated dogs as walking larders to be filled up during flush times by feeding them excess food and then harvesting them when protein was in short supply. The Aztecs developed a hairless breed expressly for eating, and dogmeat was a staple among many North American Indian tribes. Though its consumption was outlawed in 1998, dog is still on the menu in parts of the Philippines, and in Africa dogs are sometimes castrated and fattened before slaughter to encourage plumpness. You would not want to be a dog in the Congo Basin where dogs are slowly beaten to death in order to tenderize their flesh.

Dogmeat is particularly popular in Asia where approximately 16 million dogs and 4 million cats are consumed each year. The Cambridge University anthrozoologist Anthony Podberscek has studied the Asian trade in dog products. The Chinese eat more dogs than anyone else. Puppy hams are the preferred cut. Dogmeat is about as expensive as beef. The retail price for a pound of fresh dogmeat in 2004 was $2.00. Organs are a bargain: Dog brains go for about a dollar each and a penis for $1.45. Historically, the meat dog of choice in China has been the chow chow. In the 1990s, however, dog farmers (or should they be called ranchers?) decided to produce a faster-growing animal with better meat. After experimenting with Great Danes, Newfoundlands, and Tibetan mastiffs, they chose Saint Bernards as the best breed stock because of their good temperament and ability to pump out large litters of fast-growing puppies. But because Saint Bernard flesh tends to be bland, they are usually crossed with local breeds to produce tastier meat. Meat-dog puppies are harvested at six months of age when they are still tender.

South Koreans also have a long tradition of eating dog. In Korea, as in China, dog flesh is believed to have medicinal properties. Unlike China, where dogmeat is usually eaten in the winter, Koreans consider dog summertime fare. In spite of its status as a traditional food, dogmeat has become controversial in South Korea. The per capita consumption of dogmeat in South Korea is only about eight ounces a year, but with a population of 50 million, the numbers add up. According to the Ministry of Agriculture, South Koreans ate about 12,000 tons of dogmeat in 1997. In 2002, the National Dog Meat Restaurant Association was organized to promote the consumption of dogmeat and related products. These include dogmeat bread, dogmeat cookies, dogmeat mayonnaise, dogmeat ketchup, dogmeat vinegar, and dogmeat hamburger. You can also buy packs of "digested dogmeat." (I am not sure what this is.) A medicinal tonic called *gaesoju* that is said to be good for rheumatism is also produced from dogs. You don't want to know how it is made.

While South Koreans eat about a million dogs a year, more South Koreans are bringing dogs into their homes as pets. The cute little breeds—Maltese, shi tzus, and Yorkshire terriers—are particularly popular. As a result, South Koreans are increasingly ambivalent about eating dogs, and a recent poll found that 55% of adults disapproved of eating canine flesh. That said, the same survey reported that fewer than 25% of South Koreans favor a ban on dogmeat.

Taboos on eating dogs stem from two opposing facets of the human-animal relationship: Humans don't eat animals they despise and they don't eat animals they dote on. The never-eat-despicable-animals principle explains why dog eating is uncommon in India and most of the Middle East. In classical Hinduism, dogs were the outcasts of the animal world. They were despised because they were said to have sex with their own family members and eat vomit, feces, and corpses. Dogs were likened to the people on the lowest rung of the caste system, and Brahmins thought that a dog could pollute food just by looking at it. Most interpretations of Islamic law also regard dogs as unclean. Muslims, for example, are not supposed to pray immediately after being touched by a dog. Hindus and Muslims don't eat dog for the same reasons that Americans don't eat rats—they are vermin.

Americans and Europeans, however, don't consume dog flesh for exactly the opposite reason. Dogs in American households are not animals—they are family members. And because family members are people, eating a dog is tantamount to cannibalism.

What about cultures in which dogs can be either family or food? These societies usually have mechanisms that resolve the potential conflation of categories. The preferred breed of eating dog in South Korea is the *nureongi,* midsized dogs with yellow fur. *Nureongi* are not pets, and in markets in which both pet dogs and *nureongi* are sold, the pets are physically separated from the meat dogs and housed in different colored cages. The Oglala Indians of the Pine Ridge Reservation in South Dakota eat dog stew as part of religious rituals, and they also keep pet dogs in their homes. The fate of each puppy in a litter is decided soon after it is born. The pets are named; their stew-meat siblings are not.

MEAT IS DEAD AND DISGUSTING

Another factor that clouds the human-meat relationship is the guilt that comes from taking the life of another creature. Ceremonies in which hunters atone for killing animals are nearly universal in tribal societies. Most Americans repress any meat-related guilt they might experience simply by not thinking about where their dinner comes from. I successfully avoided the moral consequences of my diet until, at the age of thirty-six, I found myself, skinning knife in hand, hacking away at the steaming body of a 1,300-pound steer.

At the time, we were living on the campus of Warren Wilson College near Asheville. The college maintained a farm that included a herd of beef cattle, about thirty of which were slaughtered each year. These animals led an idyllic pastoral life and had the painless death that even animal liberation philosopher Peter Singer might have few objections to. They were never crowded into feedlots or jammed into tractor-trailers or subjected to the trauma of an industrial slaughterhouse. No, Warren Wilson cows were doted on from birth to death by the earnest back-to-the-land students who worked on the college farm crew. And, on the morning it was slaughtered, each steer was persuaded with a handful of sweet grass to walk into a small abattoir where it was shot in the head before it could say moo.

Some of the students on the farm crew knew that I studied the psychology of human-animal interactions, and one afternoon they suggested that I help them slaughter cattle the next day. After hemming and hawing a bit, I reluctantly agreed. I didn't sleep much that night. I showed up at the abattoir at seven the next morning, and an hour later I was up to my elbows in entrails. I spent the next two days helping convert big animals into packages of chilled meat.

Here's how the first steer went down. Sandy McGee, one of the students, led the animal into the kill room and tied its halter to a ring on the floor. Ernst Laursen, the farm manager, walked into the room, shot the steer with a .22, and the crew went to work. They knew exactly what to do. One of them slit the steer's throat, bleeding out the carcass, while another sawed off the animal's head and hooves. They attached a chain to the steer's legs and hoisted its carcass toward the ceiling. Out of nowhere, a wheelbarrow appeared—I did not have a clue what for. Then, using a six-inch skinning knife, Sandy made a quick vertical incision, opening the animal from rib cage to anus. Gallons of viscera plopped overflowing into the wheelbarrow. The USDA inspector examined the heart, liver, and kidneys, and stamped the carcass fit to eat.

You sometimes hear that if we had to kill our own meat, everyone would be a vegetarian. The students on the farm crew offered an opportunity to test this theory. Most of them were raised in middle-class suburbs and had never been around a cow or hog until they came to college. To assess the validity of the slaughter-leads-to-meat-rejection hypothesis, Sandy and I distributed questionnaires to the students who had participated in the killing and butchering. I also interviewed most of them.

Our results refuted the idea that slaughtering in and of itself creates vegetarians: None of the student slaughterers gave up meat. Their responses to slaughtering were, however, complicated. While nearly all of them said they found the killing and butchering process an interesting and valuable experience, most of them admitted that they sometimes felt nauseous during or after butchering a cow. Half of the farm crew said they sometimes avoided eating meat for a day or two after they had slaughtered a cow or pig, but most of them felt they had benefited from the experience. Some of their reasons were mundane. For instance, a few said they enjoyed learning about different cuts of meat and felt it would make them better shoppers. A couple of pre–veterinary medicine majors said that butchering had helped them learn anatomy. But for others, the experience had deeper meaning. It was an exercise in values clarification. They had learned where meat comes from. That it is dead.

Food psychologist Paul Rozin believes that animals and death are intimately connected in the human psyche. Rozin believes that people find many animal products, including their flesh, disgusting because animals are an uncomfortable reminder of our mortality. He writes, "Humans must eat, excrete, and have sex, just like animals. Each culture prescribes the proper way to perform these actions—by, for example, placing most animals off-limits as potential foods, and all animals and most people off-limits as potential sexual partners. Furthermore, we humans are like animals in having frail body envelopes that, when breached, reveal blood and soft viscera that display our commonalities with animals. Human bodies, like animal bodies, die."

Do people actually find animal flesh disgusting? Increasingly, the answer is yes. For instance, Scandinavian market researchers have found that the redder and more "animalized" a cut of meat is—the more it resembles a carcass—the more it turns the average consumer off. These findings pose a conundrum for the meat industry. Usually, shoppers are drawn to foods that appear fresh, juicy, and natural. But in the case of meat, fresh, juicy, and natural is seen as gross, particularly to women. The researchers recommended that the meat industry develop products that look less fleshlike—small, ready-to-cook cuts that are marinated to mask the meat's color; in short, meat that is less disgusting.

Chicken producers caught on to this a long time ago. In 1962, almost all the chickens sold in the United States were purchased as an intact carcass with heart, liver, and gizzard tucked neatly into the body cavity. You had to hack them up yourself. Not anymore. Today, fewer than 10% of chickens sold in supermarkets bear any resemblance to the body of an animal. The fastest growing segment of the retail chicken industry is officially referred to as "further processed"—translucent boneless pieces of flesh that look as if they had been grown in a Petri dish and labeled something like "tenders" or, my favorite oxymoron, "chicken fingers."

IF MEAT IS SO DISGUSTING, WHY ARE THERE SO FEW VEGETARIANS?

The process by which neutral preference comes to be regarded as immoral is called *moralization*. Attitudes toward slavery have been moralized; and more recently cigarette smoking. You would think that meat would be easy to moralize. The floor of my office is piled with books telling me why I should not eat animals. The case against meat comes down to four claims that are hard to dispute. First, to eat an animal, you have to take its life. Second, the conditions under which nearly all meat animals are raised, transported, and slaughtered involve great suffering for the animals and horrible conditions for the people who do the dirty work. Third, the conversion of plants to meat is inefficient and environmentally destructive. Fourth, eating animals causes obesity, cancer, and heart disease. Add the *yuck* factor to the moral and health arguments against meat, and you would think it would be easy to convince people not to eat flesh. But you would be wrong. The campaign to moralize meat has largely been a failure.

These days, it is fashionable—especially among young people—to claim "I don't eat meat." In a recent survey, 30% of college students said that having a vegan option at every meal was important to them, and sales of faux "meats" in the United States are growing at a rate of 35% a year. There is, however, little evidence that a wave of vegetarianism is sweeping across America. The best estimate of the number of vegetarians in the United States comes from a series of surveys commissioned over the last fifteen years by the Vegetarian Resource Group. They asked random samples of adults what foods they actually ate. These polls consistently show that between 97% and 99% of Americans sometimes eat flesh.

The animal rights movement has been successful in changing the attitudes of Americans toward the treatment of other species. But, ironically, as our collective interest in the welfare of animals has gone up, so has our desire to eat them. In 1975, when the modern animal rights movement first emerged as a legitimate social movement, the average American ate 176 pounds of meat a year. Now we are up to almost 240 pounds a year. The change in the number of meat animals slaughtered annually is even more astounding. Over the last thirty years, the number of creatures killed for human consumption jumped from 3 billion to 10 billion; from 56 animals a year for a family of four to 132 animals.

Why has the animal protection movement had so little effect on our predilection to eat other creatures? Ironically, the efforts by animal protectionists to improve the well-being of farm animals have made the consumption of flesh more, rather than less, morally palatable. Retail sales of poultry labeled organic, for example, quadrupled between 2003 and 2007.

Socially conscious consumers can now purchase meat touted as hormone-free, antibiotic-free, cruelty-free, and free-range. In other words, guilt-free. Even the supermarket in my tiny town (population 2,454) is stocked with flesh that has been given the moral thumbs-up by the American Humane Association. I can buy chickens that, I am told, led the good life: "100% all-natural," "fed an all-vegetable diet!" under "low-stress growing practices" with "no debeaking," "tunnel ventilation for fresh air," and "multiple feed bins to ensure fresh feed." The fast-food chains have jumped on the animal welfare bandwagon. McDonald's, Wendy's, KFC, and Hardee's have established high-powered animal welfare advisory boards and adopted animal care and slaughter standards.

However, the biggest reason for the jump in the number of animals killed in American slaughterhouses is the shift from eating mammals to eating birds. For many years, one of my guilty pleasures was diving into a Burger King Whopper with cheese. I loved the gooey mayonnaise sauce, iceberg lettuce, juicy fat, and charbroiled flavor I figured came from a chemical plant. My Whopper habit evaporated immediately after I read Eric Schlosser's book *Fast Food Nation*. It did not take much to convince me that Coke was as addictive as heroin, that McDonald's executives had conspired to hold down the minimum wage, and that the fast-food industry has done more to harm America's youth than Columbian drug lords. A Whopper did not taste nearly as good to me after I found out that a single ground beef patty contains bits and pieces of hundreds of cows, any one of which could be sick, and all of which spent the last weeks of their lives standing knee-deep in manure.

I was not alone in reducing my intake of steaks and burgers. Per capita beef consumption in the United States began a twenty-year slide in the 1970s when the Food and Drug Administration told us to cut down on saturated fats. The drop in our enthusiasm for beef, however, was more than compensated for by an extraordinary increase in our desire to eat chicken. While the number of cattle slaughtered each year in the United States dropped 20% between 1975 and 2009, the number of chickens killed increased 200%. The watershed year was 1990 when for the first time in history Americans ate more chicken than beef. When Herbert

Hoover ran for president on a campaign of "a chicken in every pot," the average American ate a half pound of chicken a year. Today, the figure is close to ninety pounds.

There are several reasons for our shift from eating cows to eating birds, and they have little to do with our growing concern for animal welfare. Advances in poultry science and the vertical integration of the chicken industry since the end of World War II have made chicken a better deal than beef. In 1960, a pound of chicken cost half as much as a pound of beef; now it is only a quarter of the price of beef. Beef also became linked with obesity, cardiovascular disease, and cancer. While some early claims against the hazards of red meat were based on shoddy science, recent epidemiological studies have confirmed that eating cow is bad for you. A 2009 multisite study of half a million people found that the individuals who consumed a lot of red and/or processed meat were more likely to die from cancer and cardiovascular disorders than people who ate little red meat. The authors of the report estimated that death rates of Americans would drop 11% in men and 16% in women if they ate less red meat.

Cathy Calloway, a nutritionist, tells me that her solution to these health issues is to tell her clients to eat only animals that swim or fly. But from an animal welfare perspective, the movement away from cattle has been a disaster. The average steer weighs about 1,100 pounds at slaughter, 62% of which is useable meat. A Cobb 500 broiler chicken, in contrast, yields about three pounds of meat. This means that you need to kill 221 chickens to get one cow's worth of meat. In addition, cattle enjoy longer and more pleasant lives than factory-farmed chickens. While a Cobb 500 will breathe ammonia fumes 24/7, the average steer will spend a year and half munching grass in a sunny pasture before it is hauled off to a feedlot for "finishing." A McDonald's Caesar salad with grilled chicken might be better for your health, but in terms of animal suffering, the moral scales tip toward a Big Mac.

Of course, by this logic, the food of choice for the 40% of animal activists who indicate that they eat meat would be a whale. There are 70,000 chickens worth of flesh in a single 100-ton blue whale. Ingrid Newkirk, co-founder and president of PETA, agrees. In 2001, to the dismay of some PETA supporters, the organization launched a tongue-in-cheek campaign urging people to eat whales. Newkirk explained PETA's logic to me. "We started the 'Eat the Whales' campaign to draw attention to the fact that the bigger the animal, the more meals that could be obtained from just one animal's suffering and death. In the case of whales, there is the added benefit that the animals lived free, didn't have their ears notched or their tails cut off, weren't castrated or debeaked, were never crammed into a cage that rubbed their flesh raw, never stuffed into a transport crate in all weather extremes, and so on. So, yes, to spare the greatest number of animals from suffering, if you can't (or won't) shake off that meat addiction, if you will not abandon that fleeting taste of meat for the sake of compassion and decency, your own health, and the environment, then it is better to eat body parts from the largest animal you can get your hands on."

Makes sense to me. But my daughter Betsy, who lived for a year in rural Japan, says whale meat is tough, greasy, and gross.

WHAT PEOPLE SAY VERSUS WHAT THEY DO: THE CASE OF MEAT-EATING VEGETARIANS

While the movement to moralize meat has not been particularly successful, millions of Americans call themselves vegetarians. Michele is one of them. While chewing on a piece of seared ahi tuna, she tells me that she does not eat meat. She is not unusual; most "vegetarians" in the United States eat animal flesh.

Not Che Green. Che, who was once an investment banker, is the founder and executive director of the Humane Research Council, a nonprofit organization that uses market research techniques to assess public attitudes about the treatment of animals. Like most animal protectionists, Che developed a soft spot for other creatures when he was a little kid. He ate meat as a child, though he much preferred dishes that did not remind him that he was actually eating an animal. His attitude toward meat changed in high school when he landed a summer job in an Alaskan cannery. His task was to feed the carcasses of big fish into a processing machine where they were spit out a few seconds later as canned salmon.

He made it through the summer, but the carnage got to him. Within two months he was a vegetarian and two years later, a vegan.

Che knows a lot about what Americans eat. He has collected every national survey on rates of vegetarianism in the United States. His data illustrate a fundamental principle of human psychology—what people say is often different from what they do. For instance, in 2002, *Time* reported that 6% of Americans claim they are vegetarians. However, the same article pointed out that nearly 60% of the "vegetarians" they surveyed admitted they had eaten red meat, poultry, or seafood within the last twenty-four hours. A telephone poll conducted by the Department of Agriculture also found that two-thirds of vegetarians had eaten animal flesh on the day of the survey. And one study found that teenage "vegetarians" actually eat more chicken than non-vegetarian teens do.

The Humane Research Council puts the number of true vegetarians and vegans in the United States at between 2 and 6 million. (A team of researchers from the Yale University School of Medicine concluded that fewer than one-tenth of one percent of Americans are true vegetarians.) People give up meat for a variety of reasons. Most studies have found that health concerns are the primary motivator for most vegetarians with moral/environmental concerns coming in a close second. Che's initial reason for giving up meat was first visceral and later moral disgust.

Pete Henderson's route to vegetarianism was completely different. Pete's parents were Seventh-day Adventists who did not eat meat for religious reasons, and now he is convinced of the health advantages of a plant-based diet. Unlike Che, Pete's commitment to vegetarianism has little to do with concern with the rights or suffering of other creatures. He does use a large Havahart live trap to humanely capture the animals that raid his garden.

But then he shoots them.

Pete lives on a minifarm north of Asheville, where he raises much of the food his family consumes. Five years ago, he grew tired of sharing his corn, squash, peas, beans, and blueberries with the growing population of animals that also relished fresh vegetables. He purchased a couple of live traps and would catch the animal invaders and release them a couple of miles away. When that did not work, he bought a gun. So far this year he has killed two raccoons, several turkeys, and a possum. Pete takes no plea-sure in the killing, and he is constantly trying to improve the fencing and netting around his garden so he will not have to shoot the animals raiding his veggies. But at this point, the raccoons still manage to wreak havoc on his corn, and he remains a vegetarian hunter.

Che and Pete show that vegetarians do not necessarily think alike about the morality of eating flesh. Paul Rozin and his colleagues found moral-origin vegetarians are more disgusted by meat than health-origin vegetarians, and they are more upset at the prospect of chewing and swal-lowing it. Unlike health vegetarians, moral vegetarians tend to see meat as morally contaminating and view meat-eaters as aggressive. They also tend to have more extensive rationales for eschewing flesh and reject more animal products than health-origin vegetarians. In other words, ethical vegetarians moralize meat more than health vegetarians do.

Why do some people quit eating animals while most of us manage to suppress our qualms about the morality of our food choices? According to Donna Maurer, author of the book *Vegetarianism: Movement or Moment?*, the typical vegetarian is a liberal, white, well-educated, middle- or upper-class female who is less likely than the average person to adhere to tradi-tional values. She usually gives up red meat first, and then expands her list of rejected foods to chicken and fish, and, in the case of vegans, eggs and dairy products. The motivations of vegetarians can change over the years. A person who initially gives up meat for health reasons may subse-quently internalize the moral arguments against eating animals. Similarly, vegetarians who are at first concerned with animal suffering may find that an all-plant diet makes them feel healthier.

Personality also affects whether you stop eating meat. Lauren Gold-en and I investigated the relationship between personality and attitudes toward the use of animals. Through MySpace and Facebook, we solicited animal activists, members of groups devoted to the use of animals (hunt-ers, farmers, researchers), and people not particularly concerned with animal issues to take an online survey that contained questions about their diet and beliefs about the treatment of animals. Nearly 500 people participated in the research, 40% of whom were vegetarians. Compared to the meat-eaters, the vegetarians were more creative, more imaginative, and more open to new experiences. But they were also more likely to be anxious and worried.

These findings raise an interesting issue. Most research shows that vegetarians are in better physical shape than people who eat meat, and some studies indicate that many vegetarians feel they have a better quality of life and higher psychological well-being. But is giving up meat a good idea for everyone?

MEAT AVOIDANCE AND EATING DISORDERS: THE DARK SIDE OF VEGETARIANISM

I don't know if Rory Freedman and Kim Barnouin are bitches, but they do look skinny. These two Los Angeles fashion-industry types (Rory was an agent, Kim a model) went vegan, and in 2005 wrote *Skinny Bitch,* an irreverent diet book that became a *New York Times* bestseller. With catchy chapter titles like "Sugar Is the Devil, and "The Dead, Rotting, Decomposing Flesh Diet," the book and its sequels became a media phenomenon. *Skinny Bitch* is aimed at teenage girls and young women who want to look like the authors, who are gorgeous. The book starts by asking, "Are you sick and tired of being fat?" If the answer is yes—which it usually is for young women in America—they have the answer: Stop Eating Animals.

Staci Giani, the vegetarian I interviewed who now eats raw liver for breakfast, thinks the *Skinny Bitch* message to "eat plants, lose weight" is dangerous for adolescent girls. She remembers giving up meat as a seventeen-year-old with a distorted body image. "Being a vegetarian was a way for me to have more control over my body by taking the fat out of my diet. Fat was the big evil. Emotionally, I was in a tough position in my life at seventeen. Vegetarianism gave me something to hold on to. There was something appealing about the righteousness of vegetarianism. At that age, you want to have something that is strong and clear and righteous."

Staci had opened up a can of worms. A Harris poll reported that vegetarianism is most common among teenage girls, the group that is also most susceptible to eating disorders. I had surveyed and interviewed many vegetarians over the years, but never considered that there could be a dark side to what I had always thought of as a healthy lifestyle. Could the *Skinny Bitch* approach to weight loss actually be promoting eating disorders? I immediately started looking for research on the connection between vegetarianism and eating disorders. I was surprised by what I found.

While the Skinny Bitches dangle dreams of svelte bodies in front of young women who stop eating animals, there is no guarantee that an all-plant diet will make you thin. In an article published in the *Journal of the American Dietetic Association*, "Self-Reported Vegetarianism May Be a Marker for College Women at Risk for Disordered Eating," Sheree Klopp and her colleagues found that female college students who were vegetarian did not weigh any less or have lower body mass indexes than students who ate meat. The vegetarians were, however, more likely to feel guilty after they ate and be more preoccupied with thinness. They also more frequently used laxatives and extreme exercise to lose weight, and more of them had the impulse to vomit after meals.

Other studies have reported similar findings. University of Minnesota researchers reported that vegetarian teens are almost twice as likely as their meat-eating peers to diet frequently, four times more likely to self-induce vomiting, and eight times more likely to use laxatives to lose weight. A 2009 study found that teenage and young adult vegetarians are nearly four times as likely as omnivores to engage in binge eating. (This was one of Staci's symptoms.) In separate studies, researchers in Turkey and Australia found that vegetarian adolescents are more concerned about their appearance than non-vegetarians and engage in more extreme forms of dieting. Marjaana Lindeman of the Department of Psychology at the University of Helsinki agrees that teenage vegetarianism is sometimes symptomatic of underlying emotional issues. She has found that vegetarian women, in addition to having high rates of eating disorders, also have more symptoms of depression, lower self-esteem, and more negative worldviews than non-vegetarians.

Finally, University of Pennsylvania researchers found that among college students, meat-avoiders are more obsessed with their weight, diet more often, and binge and purge more than meat-eaters. Perhaps their saddest finding was that vegetarians were much more likely than omnivores to agree with the statement, "If given the opportunity to eliminate all my nutritional needs safely and cheaply by taking a pill, I would."

My colleague Candace Boan-Lenzo studies eating disorders in young women. She has not eaten meat in fifteen years. I asked her if she knew about the evidence linking meat-avoidance to eating disorders.

"Oh, yes." she said. "I tell my graduate students about it every semester."

"What do they say?" I asked.

"They don't believe me."

Candace gets to the heart of the issue. "Vegetarianism does not *cause* people to become anorexic or bulimic. But some people, particularly teenage girls with these tendencies, use vegetarianism as a way to cover up their eating disorders. They may not even be aware of what they are doing."

Candace is right. For most people, vegetarianism offers a healthy lifestyle. Indeed research has shown that a plant-based diet is generally better for you than a diet that includes a lot of meat. I am not suggesting that most vegetarians are borderline anorexics, but we can't ignore the half-dozen studies that link vegetarianism with disturbed eating behavior, particularly in teenage girls. As with the controversial connection between childhood animal abuse and adult criminality, the important questions are how strong the link is and why it exists.

Eating disorders are serious. Bulimia, anorexia, and binge-eating disorder affect 7 million women and a million men in the United States. With a fatality rate between 5 and 10%, anorexia nervosa is among the most dangerous of mental illnesses. Clearly, more research is needed, but the *Skinny Bitch* admonition to women that *vegetarianism* = *healthy* = *skinny* is dead wrong.

WHY DO MOST VEGETARIANS
RETURN TO EATING MEAT?

Staci overcame her eating disorder. And now she eats raw meat every day. Like Staci most vegetarians return to meat. According to a 2005 survey by CBS News, there are three times more ex-vegetarians than current vegetarians in the United States. Perhaps because I was raised a Southern Baptist, I have always been fascinated by backsliders—people who have seen the light but then have a change of heart. I ran the idea of studying ex-vegetarians by Morgan Childers, an honors student who came into my office one afternoon looking for a research project. We designed an online survey and Morgan recruited participants by sending announcements of the study to MySpace and Facebook interest groups.

Within a couple of weeks, seventy-seven former vegetarians had completed our questionnaire. On average, they had been vegetarian for nearly ten years before they resumed eating meat. In his book *The Face on Your Plate: The Truth about Food*, Jeffrey Moussaieff Masson extols the health benefits of avoiding animal-based foods. He writes, "Vegetarian for most of my life, I have never really experienced illness. Now at sixty-eight, several years a vegan, I find that I have never been healthier: I weigh less than I did at thirty; I am stronger than when I was forty; I have fewer colds or minor illnesses than at fifty; and in my entire life I have experienced no major illness of any kind."

Jeffrey Masson is lucky. Not everyone thrives on an all plant diet. In our study, the most common reason vegetarians gave for resuming meat consumption was declining health. Recall that Staci reverted to omnivory because she always felt sick. Many of the ex-vegetarians that took our survey said the same thing. One wrote, "I was very weak and sickly. I felt horrible even though I ate a good variety of foods like PETA said to." Another said, "I was very ill despite having regular iron injections and vitamin supplements. My doctor recommended that I eat some form of meat as I was not getting any better. I thought it would be hypocritical of me to just eat chicken or fish as they are just as much an animal as a cow or pig. So

I went from no meat to all meat." The most succinct response was from a person who wrote, "I will take a dead cow over anemia any time."

There are other reasons vegetarians have for returning to meat. Many of the participants in our study simply grew tired of the hassle of vegetarian or veganism—they could not find quality organic vegetables locally or at a price they could afford, they did not have the time to prepare vegetarian meals, or they simply grew tired of the lifestyle. In describing the dietary difficulties he faced, the philosopher Gary Steiner wrote, "You just haven't lived until you've tried to function as a strict vegan in a meat-crazed society. What were once the most straightforward activities become a constant ordeal."

Some vegetarians simply miss the flavor of flesh. Some of our participants talked about protein cravings or how the smell of sizzling bacon would drive them crazy. One wrote, "I just felt hungry and that hunger would not be satisfied unless I ate some meat." Another was succinct. "Starving college student + First night back home with folks + Fifty or so blazin' buffalo wings waiting in the kitchen = Surrender."

MEAT AS THE BATTLEGROUND
BETWEEN MIND AND BODY

When he was a graduate student, the psychologist Jonathan Haidt read Peter Singer's book *Practical Ethics* and immediately decided the industrial production of meat was immoral. His new awareness of the cruelty inherent in industrial agriculture, however, had no effect on his diet. He writes, "Since that day I have been morally opposed to all forms of factory farming. Morally opposed, but not behaviorally opposed. I love the taste of meat, and the only thing that changed after reading Singer is that I thought about my hypocrisy each time I ordered a hamburger."

My experience is similar to Jon's. I grew up in a meat-and-potatoes family and often ate meat three times a day, usually foods that start with the letter "b"—bacon, baloney, beef. No longer. More because we like the flavors than anything else, Mary Jean and I are drawn to Mediterranean-style cuisines that are supposed to be good for you—dishes that taste of tomato, lemon, and garlic; pasta and rice entrees. We do eat meat, though much less than we used to, and usually creatures that swim or fly.

I also make what are probably symbolic gestures to reduce the cruelty of the fork. I get eggs from my friend Lydia who dotes on her mixed flock of Araucanas and Barred Rocks. I pay three times as much as I need to for chickens from Bell and Evans, whose Web site claims their chickens were "allowed to bask in the warm light of the sun." And my occasional flatiron steak comes from a Niman Ranch steer that I am told was "humanely raised on sustainable U.S. family farms and ranches." I know, however, that according to *Consumer Reports*, terms like "natural" and "cruelty-free" are usually marketing ploys that mean little.

Meat inhabits the psychological territory that Al Pacino's character in *The Devil's Advocate* called the "no-man's land in the battle between mind and body." The most natural of human interactions with animals is our desire to eat them. Meat hunger is metaphorically "in our genes," just as it is in chimpanzee genes. But though people like Jon Haidt and I cave when it comes to matters of the flesh, humans are the only species with the ability to look into the eye of a chicken and decide it would be wrong to eat it.

The Harvard University primatologist Marc Hauser, author of the book *Wild Minds: What Animals Really Think,* has remarked that the cognitive chasm between humans and chimpanzees is greater than the gap between an ape and a worm. Nowhere is the difference between humans and other animals more apparent than in matters of food. Chimps can recognize themselves in mirrors, make tools, coordinate group hunts, use symbols to communicate, and establish political alliances. But no chimpanzee has ever shown the slightest sign of remorse when ripping a tasty arm off a screaming colobus monkey.

MY RAW STEAK DINNER

A month after I interviewed Staci about her transformation from vegetarian to raw meat eater, I received an email from her: *Hal, could you and Mary Jean come over for dinner Sunday? We're having steak.*

Sure, Staci. What kind of wine goes with raw beef?

A week later, I am having second thoughts, having been lectured by my son, an emergency room nurse, and his wife, a physician, about the perils of uncooked flesh. But Sunday afternoon, we drive over the mountain. Staci gives us a tour of the farm, which is in full summer bloom. Two adolescent pigs run over to us, oinking enthusiastically; they seem genuinely glad to meet us. Then it's time to eat. For Staci, Gregory, and me, dinner is raw T-bone and a lovely Greek salad. (Mary Jean opts for baked chicken breast.) The steak, which came from a steer Staci and Gregory raised, is surprisingly good. Tender, tasty, moist. My reservations disappear. I ask for seconds and even sample a slice of raw duck breast Gregory offers me.

A couple of weeks later, I get an email from Staci that captures the moral ambiguity of the human-meat relationship.

Hal,

 We just took our pigs to the butcher this morning.

 It's amazing how complex our psyches must be in order to nurture creatures every day for seven months, only to have them sent away and then come home in little freezer packages. Or sometimes to butcher them ourselves.

 I think it takes bravery, don't you?

 I think of all the millions of humans over time who have hunted and raised animals for food because that was the way you survived. But you need to make it right in your conscience. Maybe reverence helps. Maybe killing the creature yourself helps. It completes the cycle somehow. Taking responsibility is somehow the balm that soothes the horror.

      Blessings to you and Mary Jean and to our pigs.

8

The Moral Status of Mice

THE USE OF ANIMALS IN SCIENCE

> If, in evaluating a research program, the pains of a rodent count equally
> with the pains of a human, we are forced to conclude 1) that neither
> humans nor rodents possess rights, or 2) that rodents possess all the
> rights that humans possess. Both alternatives are absurd.
>
> —CARL COHEN

> Here was the human body writ small.　　　—ALLEGRA GOODMAN

My first brush with the moral complexities of animal research was in my second year of graduate school. I had been assigned to work as a lowly assistant in the laboratory of a biochemist. One of my jobs was to collect molecules from the skin surface of earthworms. The procedure involved dropping worms into 180-degree water. Two minutes later, I would remove their inert bodies from the hot water and freeze little vials of *eau de worm* for chemical analysis. I had performed this procedure several times and viewed it as just another lab chore, one that I did not enjoy, but which also caused me no moral discomfort. The worms died instantly, and, after all, they were just worms.

One morning I was asked by the lab manager to do something different. A scientist from the another university who was studying the skin chemistry of desert creatures had arranged for some of his analysis to be done in our laboratory. Several days later, a box stamped "Caution: Contains Live Animals" was delivered to the lab. Inside was a virtual menagerie: a dozen crickets, a pair of eerily pale scorpions, a lizard about six inches long, a small snake, and a lovely little gray deer mouse. The task of liquefying the animals fell to me.

I had plunged an occasional lobster into a pot of boiling water with only a slight moral twinge, and I did not expect to be bothered by my morning's task. I lit the Bunsen burner and started to work my way up the phylogenetic scale. Like worms, the crickets died almost immediately when they hit the near-boiling water. No problem. Next, the arthropods. In the few days they had been in the lab, I had come to like the scorpions. They had an air of menace I found fascinating. They also had more body mass than the insects and took a little longer to die when I dropped them in the beaker of water. I began to wonder what I was doing.

The lizard was a striped juvenile of the genus *Cnemdopherous*. My stomach turned as I lifted it from the cage, and I began to sweat. My hands shook a little when I dropped it in the near-boiling water. The lizard did not die quickly. It thrashed for maybe ten seconds before becoming still. The little snake was an elegant racer with big black eyes. More shaking hands and sweating brow, and the thrashing reptile soon was reduced to molecules swirling in solution.

Finally, the mouse. I weighed the mouse, calculated the appropriate amount of distilled water, poured it into the beaker and turned on the heat. As the water approached the 180 degree mark, I realized that I just could not "do" the mouse. I turned off the Bunsen burner and with a mixture of trepidation and relief, walked into the office of the lab manager. I told him that I had made extracts from most of the animals but that I just could not drop a live mouse into scalding hot water. My boss did the mouse while I waited in the next room.

I have thought about my predicament many times since. In hindsight, I am struck by the similarity between my tasks that morning and the plight of the subjects in Stanley Milgram's infamous obedience experiments. As all introductory psychology students learn, the hapless participants in these studies were instructed to administer a series of electrical shocks of increasing intensity to subjects in an adjacent room. The majority of people in the experiment administered shocks they thought would be extremely painful if not lethal. Like Milgram's participants, I was confronted with a series of escalating choice points, but in my case, they were based on the phylogenetic scale rather than electric shock levels. The difference was that in the Milgram study the shocks were a ruse; the supposedly shocked "subject" was really a confederate of the experimenter. In my laboratory, the animals really died. When I look back on the incident, I get some satisfaction in knowing that I refused to boil a living mouse. But I really wish I had quit between the cricket and the scorpion.

This event provoked me to ask myself questions that I still struggle with. What is the difference between researchers who kill mice because they are trying to discover a new treatment for breast cancer and the legions of good people who smash the spines of the mice in their kitchen with snap traps or slowly poison them with d-Con? Why was it easy for me to plunge crickets into hot water, harder for me kill a lizard, and impossible for me to boil the mouse? Was it a matter of size, phylogenetic status, nervous system development, the grisly manner of their death, or the fact that the mouse was really cute? Were the results of this experiment really worth the deaths and suffering of the animals? Are they ever?

DARWIN'S MORAL LEGACY

I was not alone in my ambivalence about animal research. Even Charles Darwin struggled with vivisection—the nineteenth-century term for invasive animal research. Because Darwin was fascinated by animals, he was confronted by a problem that many modern zoologists face—sometimes you wind up killing the very creatures you have dedicated your life to studying. Jim Costa, a Darwin historian, told me that as a fledgling naturalist, Darwin shot and poisoned thousands of animals, including mice, for his collections. He was horrified by some of his own studies. He wrote of his pigeons, "I love them to the extent that I cannot bear to skin and skeletonise them. I have done the black deed and murdered the angelic little Fan-tail Pointer at 10 days old."

In the 1870s, the war on animal research heated up in England, and advocates on both sides of the issue sought the support of their country's most renowned scientist. Darwin, however, gave mixed messages. He once referred to physiology as "one of the greatest of sciences." Yet Darwin complained to a friend that surgery on animals should never be performed "for mere damnable and detestable curiosity."

Ultimately, however, Darwin sided with his fellow biologists. His views on the value of animal research are reflected by a subtle change he made in the second edition of *The Descent of Man*. In the first edition, he wrote, "Everyone has heard of the dog suffering under vivisection, who licked the hand of the operator; this man, unless he had a heart of stone, must have felt remorse to the last hour of his life." Three years later, however, he amended the sentence, adding: *"unless the operation was fully justified by the increase in our knowledge."* In 1881, he laid his cards on the table, writing in a letter to the London *Times,* "I feel the deepest conviction that he who retards the progress of physiology commits a crime against mankind."

Although Darwin put his weight behind animal research, it was his theory of evolution that muddied the moral waters by undermining the views of the seventeenth-century French philosopher René Descartes. Descartes believed that animals are biological robots and their behaviors mere reflexes. Thus scientists can slash and burn as they wish. This perspective was exemplified by the nineteenth-century French physiologist Claude Bernard, who wrote, "The physiologist is not an ordinary man: he is a scientist, possessed and absorbed by the scientific idea he pursues. He does not hear the cries of animals, he does not see their flowing blood, he sees nothing but his idea, and is aware of nothing but an organism that conceals from him the problem he is trying to resolve."

Darwin pointed out that if humans and other animals are similar in our anatomy and physiology, we also share similar mental experiences. Most modern ethologists agree. The list of psychological traits that other species share with humans is growing. Scientists have reported that elephants grieve their dead, monkeys perceive injustice, and cockatoos like to dance to the music of the Backstreet Boys. The ethical consequences of Darwin's notion that the mental capacities of humans and animals differ by degree rather than kind are inescapable. If animals have perceptions, memories, emotions, and intentions, if they can feel pain and suffer, if they dance, how can we justify using chimpanzees or dogs or even mice in experiments? Is it simply a matter of might makes right?

Thus animal researchers face a conundrum. Often, the more similar a species is to humans, the more useful it is as model for our afflictions. Because chimpanzees share about 98% of their genes with humans, they offer a better model for some human disorders than mice do. But because chimps are so similar to us, their use in research is especially problematic. In other words, often the more justified the use of a species is scientifically, the less justified is its use morally. This is the paradox of Darwin's legacy.

Animal activists sometimes claim that modern scientists are no different than their eighteenth-century counterparts in believing that animals do not feel pain. For example, in his book *Dominion: The Power of Man, the Suffering of Animals, and the Call to Mercy,* Matthew Scully, who served as special assistant to former president George W. Bush, writes, "It remains the working assumption of many if not most animal researchers that their subjects do not experience conscious pain or, for that matter, conscious anything else." Scully is wrong. For an article I was writing on animal consciousness, I once asked fourteen animal researchers if they thought mice were capable of experiencing pain and suffering. All of them said yes when it came to pain, and twelve felt that mice could suffer. In a more systematic survey of British scientists, all but two of 155 animal researchers said that animals experienced pain.

Because most animal researchers do not view animals as biological robots, they do not get off the ethical hook as easily as their nineteenth-century predecessors did. My friend Phil is an example. He studies how cells make use of glucose and fatty acids, the fuels they need to do their jobs. Phil is a basic researcher, but he hopes that his studies might someday lead to treatments for metabolic disorders such as diabetes. I asked

Phil if he ever felt guilty about using mice for his experiments. Only once, he said.

Phil had been a member of a research team that was using knockout mice to discover how cells use energy. Knockout animals are genetically engineered so that some of their genes are turned off. Phil's group was using a knockout line of mice to show that a protein called a transporter helped fatty acids and glucose cross into muscle cells, where they could be used as fuel. Because the transporter gene was inactivated in the knockout mice, the researchers predicted that they would tire more quickly than normal animals.

Phil was charged with measuring how long it took the mice to run out of gas. One way to measure fatigue in rodents is to see how long they can swim. The problem is that air gets trapped in a mouse's fur so they can float around forever, like a kid lying on an air mattress in a pool. "You have to make them swim for their lives," Phil told me. The solution is to rig up a miniature harness with just enough weight so that the mouse has to swim to keep his head above water.

Phil learned the procedure from a researcher who worked in another lab. First, take a four-inch-diameter graduated cylinder and fill it with water to within a couple of inches from the top. Then strap the mouse in the weighted harness, lower him into the water, and start the timer. After swimming a few minutes, the mouse will begin to tire. He will start to sink but then he will fight his way to the surface for a gulp of air. The trick is to let the test continue until it is clear that the animal is going down for good. Then quickly grab the beaker and dump out the water before the mouse drowns. The guy who taught Phil this procedure admitted that a couple of his animals did not make it.

Phil only tested one mouse.

He told me, "At some point I could tell that the mouse knew the score, that he had said to himself, 'OK. I know I am going to die, and I just can't do it anymore.' I was supposed to let the test continue to the point where the mouse gives up and sinks and does not try to fight anymore. I dumped the water out and the mouse just lay there panting. He was so exhausted."

Phil had had enough. He told the professor he was working for that he would not take part in the study. The task of administering the swim test was reassigned to one of the new graduate students.

Like most scientists who use mice as models of fundamental biological processes, Phil neither likes nor dislikes them. They just happened to be the best animals for him to use in order to learn how muscle cells operate. Phil has killed a lot of mice over the years with no remorse. Some by cervical dislocation (he holds their heads down with the blunt side of a pair of heavy scissors and breaks their necks by yanking their bodies backward), others by decapitation (there was a mouse guillotine in lab he worked in; it looks like a miniature paper cutter).

But when push came to shove, Phil was no Cartesian. He looked a drowning mouse in the eye and saw a creature with a will to live. "The part that bothered me was that the mouse had given up, that the mouse knew it was going to die. I would have loved to be able to do the experiment, to measure their muscle fatigue. But I could not do it. I didn't want to test their will."

THE MORAL STATUS OF SPACE ALIENS AND HANDICAPPED INFANTS: THE PROBLEM OF MARGINAL CASES

While most scientists do not deny that mice are sentient beings, I expect that most animal researchers don't spend much time fretting over the morality of their work. But every now and then, something turns your head around. In my case, it was a space alien.

It happened one rainy afternoon when our five-year-old twin daughters were bored and starting to get whiny. To placate them, I drove to the video store and rented the movie *E.T.: The Extraterrestrial,* Steven Spielberg's 1982 film about a space alien who becomes stranded in a California suburb. I figured it was just the ticket to keep them occupied for a couple of hours so I could finish writing an article about some experiments I had recently completed on snake behavior. They were immediately hooked and so was I. I quit working on my research report and watched the movie with them, not knowing that it would it would change the way I thought about the use of animals in science.

You probably know the plot. For most of the film, E.T., who has huge puppy eyes and a heart that glows, runs around Southern California with his new human pal, a boy named Elliott. The film ends when E.T.'s mom shows up to fetch her errant son. In the final scene, Elliott reaches out to E.T., pleading, "Stay?" E.T. wistfully shakes his monstrous head, looks deeply into Elliott's eyes and croaks, "Come?" But, alas, they both know it is not to be. As E.T. creeps into the flying saucer for the long ride back to the home planet Zork, Elliott blinks back a tear. So do I.

I could not get the movie out of my head. That evening over dinner, I conjured up a perverse new ending that I tried out on Betsy and Katie. I asked them, What if the movie ended differently? E.T. asks Elliott to come back to the home planet with him, and just like in the film, Elliott says no. The extraterrestrial, however, does not take no for answer. Instead, he grabs Elliott by the arm and drags the boy kicking and screaming into the alien spaceship. The doors close and as the movie ends, you hear Elliott shouting "Mommy, help me!" as the ship zooms off into space.

The reason for Elliott's abduction, I explain to the girls, is that a fatal disease is ravishing the population of Zork. Their scientists have come up with a potential cure, and humans, while not as intelligent as the Zorkians, are so biologically similar that they are good subjects for testing potential treatments. The real reason E.T. was in California was to collect humans for these important studies.

"Betsy, what do you think? Should E.T. use Elliott in painful experiments which could help save millions of Zorkians?"

"No, Daddy, no!!"

"But think about it. Zorkians are a lot smarter than humans. After all, E.T. made a space telephone out of junk, and he has special powers that we humans don't have. He could even make a dead plant bloom."

Katie chimed in, "I don't care, Daddy. It would be wrong for E.T. to put Elliott in a cage and use him for some stupid experiments."

I was not so sure. Like my daughters, I was repulsed by the specter of Elliott sitting forlornly in the alien animal colony where he is injected with an experimental drug that might save the super-smart Zorkians. But as an animal researcher, I had a problem my daughters did not share. The movie made me realize that the justification for animal experimentation, including my own research, ultimately rests on the premise that organisms with bigger brains have the right to conduct research on creatures with less developed mental capacities. Ergo, E.T. has a perfect right to haul Elliott off to Zork.

Philosophers have a different version of the E.T. dilemma that raises a similar issue. It is called the *argument from marginal cases*. Our use of animals in research is predicated on the assumption that nonhuman species lack certain abilities that humans possess—complex emotions, or abstract thinking, or the ability to learn language. But what about humans who do not possess these traits? Thousands of children are born each year with severe intellectual impairments that render them permanently incapable of ever saying a sentence or thinking about the moral status of mice. The unfortunate truth is that some people are not nearly as smart as the average chimpanzee and some humans don't have the mental capacities of a mouse. I cannot see any way to set the moral bar so it is high enough to exclude all nonhuman animals, low enough to include all human beings, and be based on morally relevant traits—the ability to feel pain counts; having two legs does not.

Would it be better to test a drug on an anencephalic infant born with no cerebral cortex, a human infant who is blind, deaf, and incapable of experiencing pain, than on a perfectly healthy mouse? My gut tells me that we should not conduct research on profoundly impaired humans in lieu of animals. But when I posed this question to the philosopher Rob Bass, he wrote back: "My gut delivers a different verdict. It seems obvious to me that research on never-to-be conscious anencephalic children is preferable to making mice suffer." Many of my students also disagree with me: They want to save the mice and conduct our biomedical experiments on death-row prisoners. That's the problem with moral intuition.

WHAT CAN WE LEARN FROM MOUSE RESEARCH?

While a few philosophers have actually argued that scientists should conduct research on severely handicapped children, most people would prefer that we use animals like mice. But supporters and opponents of animal research disagree on how much we can learn from mouse research. Geneticists say that mouse research has led to breakthroughs in organ transplantation, immunology, and our understanding of cancer and cardiovascular disorders and the causes of birth defects. They want you to know that fourteen Nobel Prizes in physiology and medicine have been awarded for studies conducted on mice. On the other hand, groups like the National Anti-Vivisection Society and the Physicians Committee for Responsible Medicine claim that studies on mice are worthless because they are hopelessly flawed and are even detrimental to human health.

Like it or not, modern biomedical research is built on the backs of mice—millions of them. As lab animals, mice have a lot going for them. They are fertile, docile, and have fast generation times (one mouse year equals thirty human years). Females become sexually mature when they are only a couple of months old and go into estrus every four or five days. They produce litters of six to eight pups after three weeks of pregnancy and will happily copulate again just two days after they give birth.

There is another reason that mice make good research animals—most people do not get in a twit about their rights. In her book *Caring: A Feminine Approach to Ethics and Moral Education*, the philosopher Nel Noddings, who believes that ethics are based on interpersonal relationships, explains why she feels no moral obligation to rodents. She writes, "I have not established, nor am I likely ever to establish, a relationship with a rat . . . I am not prepared to care for it. I feel no relation to it. I would not torture it, and I hesitate to use poisons on it for that reason, but I would shoot it cleanly if the opportunity arose." Most people feel the same way about mice. A 2009 Zogby survey found that 75% of Americans would gladly kill a mouse that showed up in their house. Only 10% indicated they would try to catch the mouse and release it outside, and no one said they would just let the mouse coexist in their home.

The transformation of the mouse from pest to model organism began in 1902 when a Harvard biologist named William Castle obtained inbred mice from a retired Boston schoolteacher for his studies of animal genetics. Castle was not the first scientist to use mice as subjects. The Austrian monk Gregor Mendel bred mice for his first tentative foray into genetics, only shifting to garden peas after his bishop deemed it unseemly for a man of God to share his living quarters with copulating animals. The laboratory mouse was officially born in 1909 when a student of Castle's named Clarence Little developed the first purebred line of lab mice. Named DBA for their coat color (*dilute brown non-agouti*), DBA mice are still used in biomedical research.

Mouse research mecca is the Jackson Laboratory, located in Bar Harbor, Maine. Founded in 1929 by Clarence Little with help from Edsel Ford (son of Henry Ford), it is a rodent factory that produces 2.5 million mice a year—nearly 40 tons of inbred, mutant, and genetically engineered mice. Scientists have their pick of over 4,000 strains of Jax mice, and, if none of them suit your needs, Jackson scientists will genetically engineer a new strain to your specifications. Making mice can be expensive, however. Developing a new strain can take a year and run $100,000. While most Jax mice are shipped out as live animals, researchers with space limitations can order their mice as flash-frozen embryos to be thawed out as needed. The names of the colors of Jax mice remind me of the muted tones on the paint-chip samples at Lowe's—"misty grey," "light chinchilla," "gunmetal."

The variety of Jax mouse infirmities is even more impressive than the colors of their fur. Hundreds of strains are afflicted with rare cancers, others are prone to facial deformities, and some are born with malfunctioning immune systems. There are Jax mouse models for defects of vision, hearing, taste, and balance. Jax mice come with high blood pressure, low blood pressure, sleep apnea, and Parkinson's, Alzheimer's, and Lou Gehrig's diseases. Researchers trying to cure infertility have their pick of eighty-eight strains of Jax mice with defective reproductive organs. Then there are the mice that just don't fit in—obsessive-compulsive, chronically depressed, addiction-prone, hyperactive, and schizophrenic mice.

Animal research advocates emphasize the successes. Liz Hodge of the Foundation of Biomedical Research tells me that without animal research we would not have immunizations for polio, mumps, measles, rubella, or hepatitis. Nor would there be antibiotics, anesthetics, blood transfusions, radiation therapy, open-heart surgery, organ transplants, insulin, cataract surgery, and medications for epilepsy, ulcers, schizophrenia, depression, bipolar disorder, or hypertension. Your pets, she says, would also suffer. We would not have immunizations against rabies, distemper, parvo, or feline leukemia, nor treatments for heartworm, brucellosis, cancer, or canine arthritis.

Mouse researchers claim that almost everything we know about the operation of mammalian genes, including human genes, is rooted in mouse studies. True, the evolutionary paths that led to mice and to men diverged 60 million years ago. My brain weighs 1,500 times more than the brain of the little fellow that lives behind the filing cabinet in my basement office. But while we have different numbers of chromosomes (he has 40; I have 46), we have roughly the same number of genes—22,000, more or less. More important, 99.9% of mouse genes have a known human counterpart.

According to Rick Woychik, president and CEO of the Jackson Laboratory, this makes mice the organism that will allow scientists to develop treatments for killers such as juvenile diabetes, breast cancer, and Alzheimer's disease. "It is," says Woychick, "a bench-to-bedside continuum. You start with basic concepts, and then these concepts mature and get translated into clinical concepts and ultimately get delivered as innovative new therapies at the bedside."

Jackson researchers are particularly enthusiastic about the new field of personalized medicine. Genes play a role in your susceptibility to nearly every disorder, from tooth decay to AIDS. Genes also affect how your body responds to medications. Some people receive no benefit from a drug but suffer serious side effects (for example, those four-hour Viagra-induced penile erections that require an immediate trip to the emergency room). Other people, however, experience few side effects and have excellent treatment results from the same medication. The goal of personalized medicine is to tell who will and who will not benefit from a drug. Woychick believes that studies based on mouse genetics will eventually enable doctors to tailor the right drug and the right dose to meet each patient's individual needs.

Carl Cohen, a University of Michigan philosopher, also believes that animal research is the key to the advancement of medicine. He is the author of a 1986 *New England Journal of Medicine* article regarded as the classic defense of animal testing. Cohen writes, "Every advance in medicine—every new drug, new operation, new therapy of any kind must sooner or later be tried on a living being for the first time. . . . The subject of that experiment, if it is not an animal, will be a human being. Prohibiting the use of live animals in biomedical research, therefore, or sharply restricting it, must result either in the blockage of much valuable research or in the replacement of animal subjects with human subjects. There are consequences—unacceptable to most reasonable persons—of not using animals in research." That's the party line, and I admit that I generally buy it—though I would like to see fewer mice killed for yet another minor variation of Claritin and instead used in the search for treatments for the neglected tropical diseases.

Opponents of animal research frame the debate differently. They throw thalidomide and Vioxx in your face as examples of the failures of tests on rodents to screen drugs that later turned out to be harmful to humans. (Mouse researchers dispute these claims.) They say scientists have exaggerated the contributions of animal research to improvements in our health. The anti-vivisectionists argue that 90% of the decline in the mortality rates for childhood killers such as scarlet fever and diphtheria came before the advent of vaccinations for these diseases. Animal research opponents argue that improvements to human well-being are really attributable to better nutrition and sanitation. They think studies on mice often lead down blind alleys and actually retard medical progress.

I support animal research and would like to dismiss the anti-vivisectionists as naïve and uninformed. However, they do have some legitimate points—including, for example, the problem of replication of research results. One reason researchers use inbred strains of mice is that they allow scientists in different labs to check each other's findings by independently confirming their results. In 1999, the world of mouse researchers was shaken up by an article that appeared in the journal *Science*. Researchers in Portland, Oregon; Edmonton, Canada; and Albany, New York, ran eight strains of mice through a series of behavioral tests using exactly the same procedures. The animals in each lab were obtained from the same sources; they were born on the same day, fed the same food, reared on the same light-dark cycle, and put through identical procedures at exactly the same age. The experimenters even wore the same brand of surgical gloves when they handled the mice.

Despite the extreme lengths the researchers took to ensure that the animals were treated alike, in some tests the mice behaved remarkably differently. A dose of cocaine completely wired the animals in the Portland lab. Their coked-up brethren in Albany and Edmonton, however, showed little response to the drug. The authors concluded that subtle differences between laboratories mean that researchers can arrive at different conclusions even when studying genetically identical animals. I filed the article in my filing cabinet under "inconvenient truth."

There is also the contentious issue of how much we can generalize from mice to humans. Biologically, there are some big differences between us and them. We live forty times longer than mice and weigh two thousand times as much. A mouse's metabolism is seven times faster than a person's. Our two species have not shared a common ancestor since the age of dinosaurs. Writing in the journal *Immunology*, Mark Davis, a professor of microbiology at the Howard Hughes Medical Institute, argued that while dozens of experimental treatments work on mice with immune system diseases, few of these have been successful on humans. He has concluded that rodents make poor models for immune disorders.

Ditto neuroscience. Amyotrophic lateral sclerosis (ALS) is a degenerative nerve disease for which there is no cure. The dead include Yankee slugger Lou Gehrig and Bob Waters, the late Western Carolina University football coach who, toward the end, was calling plays from a wheelchair, breathing through a respirator. Among its living victims are the Cambridge University theoretician Stephen Hawking. Disheartened that there are no effective treatments he could offer his ALS patients, Michael Benatar, a clinical neurologist from Emory University, read all the published mouse studies of ALS. He was surprised by the results. First, he concluded that most of the research was flawed. Often the samples were too small or the experiments poorly designed. Secondly, he found that nearly a dozen drugs that increased the life spans of mice with the rodent version of ALS had no effect when tested on humans. In fact, one drug that worked in four mouse studies made people with ALS sicker. Benatar says that using mice to study ALS is like searching for your missing keys at night under a street lamp because that's where the light is.

The anti–animal research faction, however, should not take too much comfort in the fact that some scientists are questioning the usefulness of the mouse as a model for human neurological disorders. Some neurobiologists have forsaken mice and have turned to animals whose brains are more like ours—monkeys.

HOW LABELS AFFECT OUR ATTITUDES TOWARD ANIMALS: GOOD MICE, BAD MICE, PET MICE

A recurring theme in anthrozoology is that the ways humans think about animals are mired in an uncomfortable mix of logic and emotion. Some of our decisions about the use of animals in science are perfectly reasonable. For example, an individual's attitudes about animal research depend, in part, on their perception of the potential payoff of the experiments, the degree of suffering they think the animals will experience, and the species used in the research. A survey conducted in England found that two-thirds of people approved of painful studies on mice aimed at developing a cure for childhood leukemia, but only 5% supported using monkeys to test the safety of cosmetics.

At other times our views on the moral status of animals are more convoluted. Consider the effect of labels and categories on how we think about mice. I once spent a year as a visiting scholar in the University of Tennessee Reptile Ethology Laboratory. The lab is located in the Walters Life Sciences Building, home to hyperactive marmosets, cooing White Carneau pigeons, beady-eyed albino rats, spiky green tobacco worms, and 15,000 mice. The mice were housed in spotless cedar-smelling rooms in the building's basement, where they were cared for by a competent and fully certified staff. But while all mice in the building belonged to the same species, they were not afforded the same level of moral consideration.

The vast majority of these animals were *good mice*—the subjects in the hundreds of biomedical and behavioral experiments that were conducted by faculty, post-doctoral fellows, and graduate students. Most of these projects were directly or indirectly related to the search for treatments for the various afflictions that affect our species. Though they did not have any say in the matter, these animals lived and died for our benefit. Because the university received grants from the National Institutes of Health, these mice were treated according to the Public Health Service *Guide for the Care and Use of Laboratory Animals*. Each research project involving the good mice was approved by the university animal care committee charged with weighing the costs and benefits of the experiments.

There was, however, another category of mice that inhabited the building, the *bad mice*. The bad mice were pests—free-ranging creatures you would occasionally glimpse scurrying down the long, fluorescent-lit corridors. These animals were potential threats in an environment where a premium was placed on cleanliness and in which great care was taken to prevent cross-contamination between rooms. The little outlaws had to be eliminated.

The staff of the animal facility had tried several techniques to eradicate bad mice. Snap-traps were ineffective and the staff was reluctant to use poison for fear of contaminating research animals. Thus they settled on sticky traps. Sticky traps are rodent flypaper. Each trap consists of a sheet of cardboard about a foot square, covered with a tenacious adhesive and embedded with a chemical mouse attractant—hence their other name, glue boards. In the evening, animal care technicians would place glue boards in areas where pest mice traveled, and check them the next morning. When a mouse stepped on a sticky trap, it would become profoundly stuck. As it struggled, the animal's fur would become increasingly mired in the glue. Though the traps did not contain toxins, about half of the animals were dead when they were found the next day. Mice that were still alive in the morning were immediately gassed.

Animals caught in sticky traps suffer a horrible death. I doubt that any animal care committee would approve an experiment in which a researcher requested permission to glue mice to cardboard and leave them overnight. Thus a procedure that was clearly unacceptable for a mouse labeled "subject" was permitted for a mouse labeled "pest."

This paradox was magnified when I discovered where the pest mice came from. The building, it seemed, did not have a problem with wild rodents, but in a facility housing thousands of small creatures, leakage is inevitable. Thus virtually all the bad mice were good mice that had escaped. The animal colony manager told me, "Once an animal hits the floor, it is a pest." And *poof*—its moral status evaporates.

In the Walters building, the moral status of a mouse depended on whether a creature was labeled a subject or a pest. I was quick to criticize this seemingly arbitrary distinction until I realized that the same theme was playing out in my home. For our son's seventh birthday, I kidnapped a mouse who was destined to become a meal for IM, the two-headed snake, from our lab and gave him to Adam as a birthday present. Adam named the mouse Willie and set up a home for his new pet in a cage in his bedroom. We liked Willie. He was quiet and affectionate. But mice have short life spans, and one morning Adam woke up and found Willie dead on the bottom of his cage. We held a family discussion, and the children decided a funeral ceremony would be appropriate. We put Willie in a little box and buried him in the flower garden with a piece of slate for a headstone. We stood around his grave and said a few nice things about him. Betsy and Katie cried a little; it was their first encounter with death.

A couple of days later, Mary Jean, a neatnik, discovered mouse droppings on the kitchen floor. She looked at me and said, "Kill it." That night, I put a dab of peanut butter in a snap trap on the floor between the refrigerator and the stove. I found the mouse the next morning. It was a clean kill. This time, there was no funeral. As I tossed the little guy's body into the bushes not far from Willie's grave, it struck me that the labels we assign to the animals in our lives—pest, pet, experimental subject—affect how we treat them more than the size of their brain or whether they experience happiness.

THE WASTED MICE

A woman named Susan who worked in a rodent breeding facility recently convinced me that I needed to add a new category to my typology of lab mice. In addition to good mice, bad mice, and pet mice, there are "surplus mice" that are never used in experiments. Lots of them. Some of them will be fed to snakes and owls at a zoo; most will be incinerated. Susan said that in the animal colony where she worked, surplus mouse babies were euthanized nearly every day. One of the senior techs would put handfuls of them in a clear plastic bag, insert the end of a hose that was connected to a tank of carbon dioxide, and turn on the gas. How many mice would be killed in a typical day? Susan said that it varied, but usually about fifty.

I wanted confirmation so I called John, a veterinarian I had met at a conference on the care of laboratory animals. He runs the animal colony at a major research university where many scientists use mice to figure out how genes work.

"John, I just found out that that in some animal colonies, mice are killed without ever being used for research. Is that true?"

"Sure."

"Do you guys cull surplus mice?"

"Yeah. We have a carbon dioxide chamber."

"How many animals?"

"Well, four thousand litters are born here every month. There are usually five pups per litter, so that would be a quarter of a million mice a year. We euthanize about half of them. I'd guess about 10,000 mice a month."

"Holy shit."

There are several reasons for the proliferation of surplus mice. According to Joe Bielitski, former chief veterinarian for NASA, most males are euthanized because they are prone to fight. Besides, you only need a couple of males to keep a genetic line going. He estimates that 70% of male lab mice are never used in experiments. But the most important reason for the large numbers of wasted mice is the explosion of research on genetically modified (GM) animals that began in the 1990s. Ninety percent of animal GM studies use mice. These experiments have led to important scientific breakthroughs. (A gene that affects the part of the human brain responsible for language was recently inserted into a line of GM mice. The mice did not talk, but they did squeak at a lower pitch and the gene changed the structure of their brains.) From a mouse's point of view, GM studies are terribly inefficient. It is not easy to slip a piece of DNA into the chromosome of a different species and have it be successfully incorporated into the genome. The efficiency rates of attempts to create a strain of transgenic mice range from 1% to 30%. In other words, sometimes only one animal in a hundred can be used for research. The other ninety-nine will be killed when they are a few weeks old. They are the junk mice, collateral damage.

According to calculations by Andrew Rowan, executive vice president of the Humane Society of the United States and an expert on the use of animals in research, more genetically modified mice are gassed each year in rodent production facilities than are actually used in experiments. But we actually don't know the exact number for sure because according to Congress, lab mice in the United States are not animals.

ARE MICE ANIMALS?

In 1876, the British Parliament enacted the world's first law governing the use of animals in research. The United States caught up ninety years later. The events that precipitated congressional action were a pair of articles on dogs. The first was a 1965 *Sports Illustrated* story about Pepper, a Dalmatian who disappeared from her yard one afternoon, apparently abducted by a dealer who provided animals to laboratories. Pepper's distraught owners finally located their dog, but only after she had been euthanized at the end of an experiment in a New York hospital. A year later, an article appeared in *Life* magazine titled "Concentration Camp for Dogs." Again, the story focused on the treatment of family pets who wound up as laboratory subjects. Members of the House and the Senate were bombarded with letters from constituents worried that their cats and dogs might suffer a similar fate. For a couple of months, Congress received more mail about animal research than about the two great moral issues of the time, the war in Vietnam and civil rights. The House and Senate quickly enacted the Animal Welfare Act of 1966. (It was not until 1974 that the government took steps to ensure that human research subjects were treated ethically.)

The bureaucratic gyrations of the Animal Welfare Act exemplify the convoluted ways humans think about other species. Perhaps the strangest aspect of the legislation concerns an apparently straightforward question—what is an animal? The Act's definition of the term *animal* starts reasonably enough: "Animal means any live or dead dog, cat, nonhuman primate, guinea pig, hamster, rabbit, or other such warm-blooded animal, which is being used, or is intended for use for research, teaching, testing, experimentation, or exhibition purposes, or as a pet." The smoking gun is in the next sentence. "This term excludes: birds, rats of the genus *Rattus* and mice of the genus *Mus* bred for use in research . . ."

That's right, according to Congress, mice are not animals. Nor are rats or birds. This means that 90 to 95% of the animals used in research in the United States are not covered under the main federal animal protection legislation. (Mice and other vertebrates used in research at institutions that receive grants from the National Institutes of Health are covered under a separate set of guidelines.) Federal Judge Charles Richey called the mouse/rat/bird exclusion in the Animal Welfare Act arbitrary and capricious. He was right. For instance, the congressional definition of the word *animal* means that a researcher who unobtrusively videotapes the sexual behavior of white-footed mice (genus *Peromyscus*) has to jump through all the federal legal hoops. His friend down the hall who delivers electric shocks to brain-damaged lab mice (genus *Mus*), however, is exempt from the regulations.

It is instructive to compare how the Animal Welfare Act treats mice, a species most people do not like, with our best friend, the dog. Because mice are not animals, they have no standing under the law. End of story. Dogs, in contrast, are singled out for special treatment. They are entitled to a daily dose of "positive physical contact with humans" (I think this means play). Ironically, because the act applies to dead as well as living animals, dead dogs have more legal protection than live mice. (A footnote in the Animal Welfare Act, however, exempts dead dogs from the husbandry and cage-size requirements.)

The mouse/rat/bird exclusion means that we have no idea how many animals in total are used in research each year in the United States. I can tell you that exactly 66,314 dogs, 21,367 cats, 204,809 guinea pigs, and 62,315 apes and monkeys were used in biomedical and behavioral experiments in 2006. But no one has a clue about the number of mice used in research in American laboratories. Some experts say 17 million. Others, including Larry Carbone, a lab animal veterinarian at the University of California San Francisco and author of the book *What Animals Want: Expertise and Advocacy in Laboratory Animal Welfare Policy,* say the number is much higher. Larry puts it at well over 100 million. The larger number is probably closer to the truth.

The Animal Welfare Act has been tweaked often over the years, but the most important amendments were added in 1985 when Congress took on the issue of which studies were worth doing. In Great Britain, every animal experiment must be approved by the Home Office in London. Congress took a different route and placed the responsibility for ensuring the ethical treatment of lab animals on the institutions where the research was conducted. They directed each institution to establish a local Institutional Animal Care and Use Committee, or IACUC.

Serving on an IACUC is a tough job. At major research universities, animal care committee members can spend hours each week poring over the fine print in proposals that can run fifteen or twenty pages. Every couple of months, they get together and play God. The members thrash out which proposals to approve, reject, or request more information about. The lives of animals hinge on their decisions, as do scientific careers. Being an IACUC member is a good way to lose friends. But can a committee accurately weigh the benefits of an experiment against the costs in terms of animal suffering?

JUDGING THE JUDGES: HOW GOOD ARE THE DECISIONS OF ANIMAL CARE COMMITTEES?

Some years ago, I received a phone call from Scott Plous, a social psychologist from Wesleyan University who is an expert on decision making. Both of us were interested in how people think about other species, and we had once run into each other while handing out surveys to activists at an animal rights demonstration in front of the Capitol in D.C.

"Hal, have you ever considered doing a study where you would ask different animal care committees to evaluate the same proposals?" he asked.

"Of course," I replied. After all, it would be nice to know that the system Congress set up to ensure the welfare of research animals worked, that the University of Texas Animal Care Committee and the Johns Hopkins Animal Care Committee would make the same decisions about the same experiment. "But, Scott, it would be impossible. Scientists are busy. You would never get them to cooperate."

Scott disagreed. He thought committees would happily participate if you offered them money they could use to enhance animal care at their university. I was skeptical, but said, OK, count me in. Scott pitched the idea to the National Science Foundation and they approved the proposal. He was right—by offering institutions extra funds for animal care, we easily recruited fifty randomly chosen university IACUCs to participate in our study. Indeed, the committees were enthusiastic about the project. In the end, roughly 500 scientists, veterinarians, and community members took part in the study, nearly a 90% response rate.

Each committee chairperson sent us three animal researchers' proposals their group had already reviewed. After removing identifying information, we sent them the proposals to be re-reviewed by a second committee. The research ranged from studies of how bats find water to the development of eating disorders in mice. In all, the 150 proposals involved over 50,000 animals, mostly mice and rats, but also a smattering of other species—chimpanzees, frogs, buffalo, egrets, pigeons, dolphins, monkeys, sea turtles, bears, lizards, you name it. When the data were in, I flew up to Connecticut to help Scott crunch the numbers and figure out what they meant. I had served as an animal care committee member, and I was sure that there would be fairly high levels of agreement between the first and second IACUCs. I was so wrong.

There are moments of truth in science. For me, it is the fraction of a second between the time you push the enter key on the computer and when the results flash on the screen. Scott and I sat in his office, our eyes on the monitor. I was antsy, feeling a little rush of anticipatory adrenaline, like an offensive lineman waiting to hear the quarterback yell "Hutt!"

Scott pushed the button. The numbers popped up. Our jaws dropped.

About 80% of the time, the second committees made a different decision than the first. Our statistical analysis indicated that the committees might as well have made their decisions by flipping a coin. Clearly, the system was inadequate. Why, I wondered, should it be OK to shock dogs at one university, but not at another one? In retrospect, I should not have been surprised to find that the decisions of animal care committees are wildly inconsistent. It is harder than you think to tell good from bad research. In his novel *Zen and the Art of Motorcycle Maintenance*, Robert Pirsig laid the issue out nicely: "But, if you can't say what Quality is, how do you know what it is, or how do you know that it even exists?" This is a question that can keep a scientist up at night.

Our findings that different animal care committees often make different decisions were not an anomaly. Studies showing large inconsistencies in peer-review judgments of quality in science go back thirty years. They include studies of ratings of grant proposals, journal article submissions, and the decisions of both human and animal research ethics committees. The truth is that scientists have problems discerning the quality and importance of research. This is a little secret that researchers do not like to think about.

Put simply, the system Congress enacted to oversee the treatment of research animals is fraught with inconsistencies. Why are white-footed mice but not lab mice covered by the Animal Welfare Act? Why are dogs but not cats entitled to a play session every day? Why can a project be given full approval by one animal care committee and flat-out rejected by a different one? Unfortunately, these nagging problems give credence to the charges by anti-vivisectionists that the fox is presently guarding the henhouse.

What can we do? For starters, Congress should extend the Animal Welfare Act to include all vertebrate species—mammals, birds, reptiles, amphibians, and fish. (British animal research regulations even extend to octopuses.) Our data suggest that most scientists also want mice, rats, and birds covered under the Animal Welfare Act. Three-fourths of the animal researchers who participated in our animal care committee study said they disagreed with the Animal Welfare Act's definition of the word *animal*.

Of course, we could just dump the present system. We could either let scientists conduct animal research without any external oversight or we could throw a pair of dice to decide which animal experiments should be conducted. Both choices are unacceptable. Some animal rights activists argue for a third alternative. They would have us ban animal research altogether. But people who oppose all animal experimentation are up against their own inconsistencies and paradoxes.

THE ANIMAL RESEARCH PARADOX: USING ANIMAL EXPERIMENTS TO SHOW THAT YOU SHOULD NOT CONDUCT ANIMAL EXPERIMENTS

The argument against animal research is based on the premise that mice and chimpanzees fall within the sphere of moral concern but that tomato plants and robots do not. That's because animals have mental traits that plants and machines don't possess. For example, the philosopher Tom Regan restricts the possession of rights to species that have consciousness, emotions, beliefs, desires, perceptions, memories, intentions, and a sense of the future. But how do we know which animals have these attributes? The answer, of course, is animal research.

The legal scholar Steven Wise is one of the few animal rights advocates who has seriously grappled with the moral implications of differences in mental capacities among the species. In his book *Drawing the Line: Science and the Case for Animal Rights,* Wise developed a 0 to 1.00 "autonomy scale" on which species are rated according to their cognitive abilities. The rankings are based on Wise's review of scientific studies of animal behavior and cognition. Humans are assigned a 1.0 on the scale; chimpanzees, .98; gorillas, .95; African elephants, .75; dogs, .68; and honeybees, .59. Wise argues that creatures scoring above .90 (great apes and dolphins) are entitled to basic legal rights, while animals with scores below .50 are not. The strength of this approach to animal ethics is that a species' moral standing is based on evidence of their cognitive capacities, rather than naïve conjectures about their abilities or how much we like them. For instance, after reviewing the science, Wise concluded that African gray parrots have a slightly stronger claim to basic rights than do dogs.

There is, however, a paradox associated with this empirical approach to animal rights—you need to conduct animal research to determine if it is immoral to use a species in animal research. Wise, for instance, assigns dolphins an autonomy scale score of .90, which puts them in the highest category of nonhuman creatures that deserve legal rights. He writes, "Dolphins have concepts and spontaneously understand pointing, gazing, and the holding up of replicas. They instantly imitate actions and vocalizations." His assessment of the cognitive abilities of dolphins is based on the finding of a University of Hawaii psychologist named Lou Herman. Over three decades of research, Herman has demonstrated that dolphins have extraordinary memories, understand human gestures better than chimpanzees, and have such sophisticated linguistic skills that they will correct your grammar.

Given that Wise bases his case for dolphin rights on Herman's research, you might think he would be fan of these studies. Wrong. In fact, he vehemently argues that Herman's dolphin research is unethical, that Herman exploits his animals, and that he treats them like prisoners. The irony, of course, is that without these studies of dolphin cognition, Wise would not be able to argue that the mental abilities of dolphins are comparable to those of chimps and thus dolphins are entitled to legal rights.

What about mice? Where do they fall on the autonomy scale? Wise does not mention them in his book, so I sent him an email: *Professor Wise: Where do mice rank on your scale? They are, after all, the most common mammal used in research.*

Wise replied that the omission of mice was a matter of time constraints. The autonomy rankings, he said, are based on an objective review of the available evidence for the capacities of each species. This task requires tracking down the latest research reports and interviewing leading scientists who have studied the behavior and cognitive abilities of each species. Wise said that in the cases of apes and dolphins, the data fit his preconceptions. On the other hand, honeybees scored much higher than he ever anticipated. The evaluation process takes roughly three months for each animal, but there are only twenty-four hours in a day and thousands of species.

DO MICE EXPERIENCE EMPATHY?
THE MCGILL PAIN STUDIES

Wise freely admits that we don't know enough about the minds of most species to accurately place them on his moral status scale. This would seem to mean that we need more, rather than less, animal research. Some of these studies would certainly discover that some animals have unexpected capacities. Researchers at McGill University's Pain Genetics Laboratory, for example, recently conducted a series of experiments that they think show that mice are capable of empathy. I am not convinced that mucine empathy is analogous to the human experience of empathy, but their findings do raise interesting ethical issues.

The purpose of the studies was to discover whether mice would react to pain inflicted upon other mice. The researchers used several procedures to induce pain in the animals. Most of the mice were given the unfortunately named "writhing test," in which they were injected in the stomach with a dilute solution of acetic acid. Others were injected in their hind paw with an irritating liquid, and a final group was subjected to the paw withdrawal test which involved measuring how quickly a mouse would lift its feet from a hot surface. If I have calculated correctly, the research involved over 800 mice.

Did the mice feel each other's pain? The short answer is yes. The animals injected with acetic acid writhed more when tested near another writhing mouse than when tested alone. But—and here is the interesting part—pain contagion only occurred when the other mouse was a relative or a cage-mate. Mice showed no signs of empathy in the presence of suffering strangers!

How does a mouse know if his cage-mate is suffering? Does he see an agonized look in his friend's eyes or hear their ultrahigh-frequency moans? Or perhaps a mouse in pain emits an odor that signals fear. The researchers checked each of these possibilities by systemically disrupting the sensory systems of mice. Vision was easy. They just put an opaque screen between two writhing mice. Eliminating the sense of smell was harder. After injecting the mice with a local anesthetic, they flooded each of their nostrils with a caustic chemical that fried their smell receptor cells. This procedure permanently destroyed their ability to ever smell anything. To eliminate hearing, they injected mice with a chemical called kanamycin every day for fourteen days. Two weeks later, the mice were permanently deaf.

From a scientific perspective, the experiment was a success. The researchers discovered that mouse empathy is the result of vision alone. The mice deprived of their senses of smell or hearing remained empathetic. The mice that were blocked from seeing their suffering compatriots were not.

Was the research ethical? Pretend for a moment that you are a member of the McGill University animal care committee charged with approving or rejecting research proposals involving animals. How would you have voted? Did the results of the experiments justify the pain and suffering of the animals?

Make your decision: approve or reject.

For me, this is a tough one. The research was well done, and while most scientific articles are never read by anyone except perhaps the author's mother, the results were published in the journal *Science* and garnered worldwide publicity. Further, the researchers made a reasonable argument that the pain was relatively mild and short-lasting.

But I vote to reject.

The reason I would not approve the study is that listening to the Rolling Stones played loud is one of life's great pleasures, and I love the toasty aroma of fresh French bread. Hence, I did not like the idea of deafening and wiping out the sense of smell of so many mice. (I would probably have approved the study if the researchers had agreed to dump the sensory deprivation experiments.)

When I read their research report, my first thought was, "These guys are in deep shit." I figured they would be getting death threats from the lunatic wing of the anti-research movement. I was completely wrong. The radical Animal Liberation Front, a group that advocates the harassment of animal researchers, prominently featured the McGill mouse pain study on their Web site as proof that humans and mice are kindred spirits. Even some scientists who normally oppose experiments that involve the infliction of suffering and permanent deformity in research animals seemed to tacitly approve of the study. Marc Bekoff is an eminent ethologist and a powerful voice for animal protection. He argues that scientists should not conduct research on animals that they would not do on their own dogs. Thus I was surprised to find that Marc used the mouse pain study in his book *Wild Justice: The Moral Lives of Animals* as evidence that even rodents experience sophisticated emotional states.

Like Marc, Jonathan Balcombe is both an animal activist and a scientist. (His doctoral research was on the behavior of bats.) The author of *Second Nature: The Inner Lives of Animals*, Jonathan opposes all invasive and painful research on other species and is a popular speaker in animal protection circles. Articulate, thoughtful, and calm, Jonathan is the perfect face for a movement that is often stereotyped as a band of wild-eyed fanatics. Given his opposition to invasive animal research, I was puzzled when Jonathan used the McGill pain study results to argue that mice have emotions in a lecture at a conference I recently attended. Jonathan is an old friend and I called him up to find out how he would have voted if he were on the McGill animal care committee.

"Of course, I would have voted to reject it," he said.

"But don't you find it ironic that so much of what we know about the mental abilities of animals is based on research that would not be permitted if you had your way and experiments on captive animals were abolished?" I asked.

Jonathan was ready for that question. It turns out that he gets asked this a lot during the Q-and-A sessions that follow his talks on university campuses. Inevitably, someone will stand up and ask, "Dr. Balcombe, you oppose animal research, yet your argument that animals are conscious beings relies on the results of experiments that have harmed animals. Isn't that a contradiction?"

This is not an ethical dilemma for Jonathan. "I hate these studies," he tells the audience. "If I had my way, we would not allow some of the research that I use to show that animals have feelings. But the fact is that they have already been done, and they do shed light on the question of animal consciousness. So I am going to keep using them. "

It is obvious to me that Jonathan has thought a lot about this issue, but I am surprised when he brings up the Nazi medical experiments in our conversation. The reason is that, prodded by the mouse pain study, I have also been thinking about them. Dr. Sigmund Rasher, a German physician, immersed prisoners in Dachau in frigid water for extended periods to see how long pilots could survive if they were downed in the icy waters of the North Sea. Dozens of his subjects died. By some accounts, this research remains some of the best information we have on the effects of hypothermia on the human body. Some medical ethicists believe that because the data derived from the experiments at Dachau and Auschwitz are already collected, we honor the dead by using the information to save human lives today—even if it was obtained unethically. Others, however, argue that the data are morally tainted, ill-gotten gains that should not be used under any circumstance. Similarly, some animal activists believe that the results of experiments on animals are also ill-gotten gains. They believe, for example, that it is immoral to take medicines that have been tested on animals.

Are the results of the McGill pain experiments or, for that matter, studies of language learning in captive chimpanzees and dolphins, also ill-gotten gains that should not be used, even to make the case against animal research? Jonathan is not losing much sleep over this one. When it comes to the campaign against animal research, he admits to me that he has reluctantly become a utilitarian. "I am willing to use any available evidence to plead the case of the animals. Whatever works," he says. But then he adds, "Within reason."

WOULD YOU KILL A MILLION MICE TO CURE DENGUE FEVER?

But reason can be elusive in the debate over animal research. I think the argument for animal research is stronger than that for any other human use of animals, including eating them. Others disagree. Indeed, public opinion polls indicate that more Americans object to animal research than disapprove of hunting.

The moral status of mice came up in a discussion I recently had with my colleague Linda, an English professor whose writings focus on inequality and oppression. She is deeply concerned with the exploitation of both animals and impoverished people, particularly people in post-colonial Africa. Linda has been involved in animal protection since she was a teenager. She and her husband are vegans. In her spare time, she volunteers at a sanctuary for farm animals. Linda does not wear leather and she hates zoos and circuses. Linda believes that animal exploitation is intimately tied to the oppression of women, minorities, and people of color.

"Animal abuse is the foundational form of oppression," she tells me.

For Linda, activities like hunting, raising animals for fur, and eating them are simple moral issues. They are wrong: end of story. But even for her, animal research is a quagmire.

"I don't believe humans have any right to use other species for our own benefit," she says. Then she adds, "On the other hand, I do think that some research might actually benefit humans."

"Can I press you on this?" I asked.

"Sure."

"What if a drug company decided to spend less money on erectile dysfunction commercials and more on research aimed at developing cures for neglected tropical diseases that destroy the lives of so many people living in developing countries? Would you be willing to sacrifice a million mice for a vaccine against Dengue fever, one of the leading causes of child mortality in sub-Saharan Africa?"

Linda looks at the floor.

After a long pause, she says, "I am not sure. I can't decide one way or another."

"Why not?" I ask.

She replies, "Well, I do not believe that the lives of human beings are more valuable than the lives of other animals. But *mice*?"

Linda's conflict about using mice to develop a vaccine that could potentially save millions of African children is understandable. Her belief in animal equality was bumping up against her commitment to improving the lives of the world's poorest people. But for me, this is not a tough call. Yes, I would swap a million mice to wipe out Dengue. In a heartbeat.

But a million mice for a treatment for baldness? Or erectile dysfunction? Hmm . . . probably not.

9

The Cats in Our Houses,
the Cows on Our Plates

ARE WE ALL HYPOCRITES?

Am I saying that a spider has as much right to life as an egret or a human? Yes. I see no logically consistent reason to say otherwise.

—JOAN DUNAYER

Stop smirking. One of the most universal pieces of advice from across cultures and eras is that we are all hypocrites, and in our condemnation of others' hypocrisy we only compound our own.   —JONATHAN HAIDT

If you visit Seattle, don't miss the Pike Place Market. Every year, 10 million visitors flock to the flower stalls, bakeries, fresh produce stands, and the assorted gourmet cheese, candy, mushroom, fruit, and salami shops. The biggest draw is the Pike Place Fish Market, where men wearing rubber boots and gray hoodies confidently fling fifteen-pound king salmon twenty feet through the air into the arms of another man in a hoodie standing by the cash register. The crowd loves to see the big fish fly. They laugh and take pictures. I have seen it myself. I laughed and took pictures, too.

In June 2009, the American Veterinary Medical Association decided that a fish-catching demonstration would be a terrific team-building exercise for the 10,000 veterinarians and para-professions attending the organization's annual meeting in Seattle. PETA was not amused. In an article in the *Los Angeles Times*, a PETA campaign manager named Ashley Byrne was quoted as saying, "Killing animals so you can toss their bodies around for amusement is just twisted. And it sends a terrible message to the public when vets call it fun to toss around the corpses of animals." The media played it for laughs, and my first thought was that Ashley needed to get a life. But then PETA issued a statement saying that the crowd in the market would not be laughing so loud if the guys in hoodies were throwing around the bodies of dead kittens. And I realized PETA was right. Why should people think it is funny to play catch with a dead fish but not a cat carcass?

OUR ATTITUDES ABOUT THE TREATMENT OF ANIMALS ARE OFTEN INCOHERENT

Elizabeth Anderson, author of the book *The Powerful Bond between People and Pets,* is troubled by this kind of moral inconsistency. She is, for example, puzzled by pet owners who wear fur coats. Anderson writes, "How a person who has ever loved or kissed a puppy or a kitten can turn a blind eye to the anal electrocution of a mink or the head-bashing of a seal pup, I doubt I will ever understand." She should not be surprised that a person who melts at the sight of a kitten can also love the luster of mink. Glaring inconsistencies in our relationships with other species are common. They occur even among many people who take the rights of animals seriously. The social psychologist Scott Plous found that 70% of animal activists who felt that the use of animals for clothing should be the top priority of the animal rights movement admitted that they wore leather products.

Psychologists have long known that our words and our deeds are often at odds. A widely accepted theory of attitudes is called the A-B-C *model*. It holds that attitudes have three components—Affect (how you feel emotionally about an issue); Behavior (how the attitude affects your actions); and Cognition (what you know about the issue). Sometimes the components work together. Rob Bass is a good example. Rob is a fifty-two-year-old philosopher whose life was going along just fine until 2001, when he came across an article by an ethicist named Mylan Engel. Engel made an argument against eating animals that Rob—to his surprise—found utterly compelling. Rob figured there had to be a flaw in Engel's logic and he spent the next three weeks trying to disprove the argument. After a month, he gave in. Once he became convinced that Engel was right (a cognitive shift), he knew he had to quit eating meat (a behavioral shift). A few weeks later he walked by the college cafeteria and caught a whiff of burgers frying on a grill. His response was immediate and visceral: "Yuck—that smell is disgusting!" (an affective shift). Engel's article had started Rob on a cycle in which his behavior, thinking, and emotions reinforced each other. Now Rob and his wife, Gayle Dean, who went through a similar transition, are strict vegans. They are opposed to the exploitation of animals of all kinds and Rob teaches animal rights in his ethics classes.

Rob and Gayle, however, are the exceptions. Most people seem blithely untroubled by the contradictions in their attitudes about animals. The *Los Angeles Times* once commissioned a survey in which a random sample of American adults was asked whether they agreed or disagreed with the statement "Animals are just like people in all important ways." The paper reported that 47% of respondents agreed with the statement. I was skeptical about the results, so I decided to see how my students responded to the item. I gave one hundred of them a survey that included the *L.A. Times* item as well as a dozen other questions related to the treatment of animals. My skepticism was unjustified. Exactly 47% of the students also agreed that animals were just like people in all important ways. But their belief that humans and animals are equal had little effect on their attitudes about the use of animals. Half of the students who said that animals were "just like people" favored the use of animals in biomedical research, 40%

of them thought it was OK to replace diseased human body parts with organs taken from animals, and 90% of them regularly dined on the creatures they believed were like humans "in all important ways."

How can people maintain such blatantly contradictory opinions? Most people's views about the treatment of other species exemplify what psychologists call "non-attitudes" or "vacuous attitudes." These are superficial collections of largely unrelated and isolated opinions, not the coherent belief system that we see in people like Rob and Gayle who have thought deeply about moral problems involving animals. The ethical issues associated with our relationships with other species are complex, and most people, even people who say they are animal lovers, are somewhere in the middle. For instance, when asked in a National Opinion Research Center survey how they felt about animal testing, only one in five adults said they had strong opinions one way or another about the topic.

While there are plenty of exceptions, most people don't get lathered up about the treatment of animals. In 2000, the Gallup Organization asked American adults to rate the importance of social issues like abortion rights, animal rights, gun control, environmentalism, women's rights, and consumer rights. Animal rights came in last. In 2001, the Humane Society of the United States commissioned a survey in which people were asked which national animal protection organization did the most to protect animals; half of the respondents could not name a single organization that promotes the interests of other species. Finally, a survey of people who boycotted consumer products reported that only 2% did so out of concern for animals. The fact is that, other than our personal pets, the treatment of animals is not particularly high on most people's list of priorities.

If you really want to know how people feel about the treatment of animals, follow the money. Americans donate between $2 billion and $3 billion dollars each year to animal protection organizations. That sounds like a lot until you compare it to the money we spend to kill animals: $167 billion on meat; $25 billion on hunting supplies, equipment, and travel; $9 billion to kill animal pests; $1.6 billion on fur clothing. Of course, we spend vastly more on the well-being of our own pets than we contribute to organizations that promote the welfare of nonhuman creatures we do not personally know. This is perfectly consistent with several fundamental principles of human nature. One is the well-established evolutionary principle that family comes first—and pets are now considered family members in many American homes.

Another is a phenomenon that the University of Oregon cognitive psychologist Paul Slovic calls "psychic numbing"—the larger the tragedy, the less people seem to care. For instance, individuals say they will donate twice as much to save one sick child as they will to save a group of eight sick children. Human indifference gets magnified in cases of mass suffering. As *New York Times* columnist Nicholas Kristof pointed out, psychic numbing helps explain why New Yorkers got very riled up about the eviction of a single red-tailed hawk from his nest on the ledge of a luxury Fifth Avenue co-op, but showed little outrage about the plight of 2 million homeless Sudanese. Slovic refers to human indifference in the face of overwhelming numbers as "the collapse of compassion."

Not everyone suffers from compassion collapse. Eleven million Americans are members of the Humane Society of the United States. The ASPCA claims a million members, PETA over 2 million. Many of these people don't just contribute money—they also take action. One of the first projects I undertook as an anthrozoologist was a series of interviews with animal activists. I focused on people working at the grassroots level, the foot soldiers—not movement leaders, philosophers, or celebrities. My purpose was to discover the types of people who are drawn to the animal rights movement, to understand why they became involved in animal protection, and how this moral commitment had affected their lives.

It turns out that three out of four animal rights activists are women, and most of them are politically liberal, well-educated, solidly middle class, and primarily white. Nearly all of them, of course, have pets. Animal activists come to their movement via different paths, but the most common thread is moral shock. For Katherine, a nurse, the shock was caused by a single photograph.

"What drew you to the animal liberation movement?" I asked her.

"A picture on a PETA poster. I can still remember the picture of that little monkey. They had severed his nerves, and he couldn't use his arm. They had taped the other arm and made him use the handicapped arm."

"You still remember what that picture looked like?"

"Oh, yes." she said. "This monkey had really beautiful eyes and it looked like it had been crying. It makes me feel like crying."

At this point Katherine did begin to cry softly, and she said, "I didn't realize I was so emotional about it until I started talking about it."

Opponents of animal rights will run into someone like Katherine and assume that that all animal activists are hypersentimental types who prefer the company of animals to people. This is a mistake. Many of the activists I have spoken with have a firm, rational basis for their opposition to the exploitation of animals. One woman who was very conversant with the nuances of the intellectual case for animal rights said she resented it when people call her "soft-hearted." She told me, "To pass off all the years I have been thinking through these issues as being 'soft-hearted' is really condescending."

ANIMAL LIBERATION AS RELIGION

As a group, animal rights activists are not very religious, at least not in a conventional sense. In one study, only 30% of marchers at a large national animal rights protest indicated that they were members of traditional religious denominations, and about half of them said they were atheists or agnostics. But as with other moral crusades, the animal liberation movement has religious elements. Like religious belief, animal activism can give meaning and purpose to a person's life. When I asked Phyllis how important the animal rights movement was to her, she blinked and seemed puzzled, as if the answer were self-evident. "It is my life."

Mark was a retired policeman who had been clinically depressed before he and his wife became involved in animal protection. He felt that the animal rights movement saved him. He told me, "It's one of those things that happen in one's lifetime that make you happy doing what you are doing," he said. "It does affect your whole existence. We are just totally happy."

You get the sense when you talk to Mark that, like Saint Paul on the road to Damascus, the scales were suddenly pulled from his eyes and he saw the light. A self-confessed agnostic named Brian said to me, "Sometimes I laugh at myself. I know how a 'born again' probably feels. Just like me, their beliefs affect every aspect of their lives." Another activist told me, "I have grown to respect Jesus in a very different way. I think that if Jesus were alive today, certainly he would be a vegetarian. I think he would be an animal rights activist."

Animal rights activists and religious fundamentalists are alike in another way—they see moral issues in terms of black and white rather than shades of gray. Shelley Galvin and I gave animal activists a psychological scale developed by the social psychologist Donelson Forsyth to assess individual differences in people's ethical ideologies. Seventy-five percent of animal activists (compared to only 25% of a group of college students) fell into the "moral absolutist" category. People with this ethical stance believe that moral principles are universal and that doing the right thing will result in happy endings.

THE CONSEQUENCES OF TAKING ANIMALS SERIOUSLY

Big things happen when you decide to take animals seriously. First, you have to change your life. All of the animal activists I have met have taken steps to bring their behavior in sync with their beliefs. Some take baby steps, others giant steps, and some are more successful than others. Marie was the biggest failure; she only lasted two weeks. During lunch at her first (and last) animal rights conference, Marie had a Big Mac attack and snuck over to a McDonalds. That was the end of animal rights for her. She was the exception. Of activists I surveyed in Washington, D.C., who were attending a large demonstration, 97% had changed their diet (though many of them still ate some meat), 94% purchased consumer products labeled "cruelty-free," 93% boycotted companies that tested products on animals, 79% said they avoided clothes made from animals, and 75% had written letters about the treatment of animals to newspapers or legislators. New beliefs and new behavior reinforce each other; a woman named Gina told me, "The more I got involved, the more my diet changed. And the more my diet changed, the more involved I got."

The moral commitment of activists shows up in many ways. Some, for example, refuse to kill animals that are normally regarded as pests. One man had recently spotted a copperhead snake in his garden. A year before, he would have grabbed a hoe and killed it; now he carefully nudged it back into the woods. Bernadette was an IBM executive who seemed to live a completely conventional upper-middle-class existence complete with husband, two kids, a minivan, and a dog. What made her different from the other women in her subdivision was that she would not kill a flea.

"Bernadette," I asked. "Can you give me an example of how your views on animal rights affect you on a day-to-day basis?"

"Well, I don't use toxic chemicals on my dog to get rid of fleas. Instead, I try to pick them off and put them outside. I know they do not feel pain or anything, but I feel it is important to be consistent. If I draw the line somewhere between fish and mollusks, it isn't going to make sense."

But then the roaches showed up. "We recently annihilated the roaches in our house," she said. "But before we resorted to Terminix, I walked around for a week trying to telepathically tell the roaches, 'You have invaded my territory and we are going to take drastic action.' In my fantasy, I was hoping they would magically disappear." They didn't.

Bernadette was up against the "activists' paradox"—*the greater your moral clarity, the harder it is to be morally consistent.* Small things can become an issue. For Gina, even eating plants posed a dilemma. She sometimes wondered if a fruit and nut diet was ethically preferable to eating plants like carrots that do not survive harvest. Roy's passion was church-league softball. After months of searching, he found a satisfactory (but not great) synthetic glove. He could not, however, find a decent ball that was not made of leather. Fortunately for Roy, a lot more products are available for people seeking a cruelty-free lifestyle. These include vegan-friendly condoms as well as synthetic softballs.

In his book *The Happiness Hypothesis,* Jonathan Haidt argues that the keys to a happy life are a sense of virtue and moral purpose, a feeling of enlightenment, volunteerism, and solidarity with a group with whom you share core values. Many animal activists have these, so you would think they would be among the happiest people on earth. This was certainly true of the woman who told me at a protest against the mistreatment of captive black bears, "I just feel sorry for people who don't have something like this in their lives."

But some activists pay a heavy price for their moral vision. For instance, their allegiance to animal liberation can alienate friends, family, and lovers. While most Americans *say* they support the notion of animal rights, people are often uncomfortable around individuals who take moral issues seriously. An activist named Alan told me, "My friendships have suffered a great deal. Nobody understands what I am doing, and I feel a lot of defensiveness from them. I lost my closest friend, and a lot of it had to do with animal rights."

Commitment to animal protection can affect a marriage. For Hugh and Lydia, the cause of animals is a joint commitment, a common focus that makes their marriage stronger. They cook vegan meals together, go to the same conferences, they discuss issues, and give each other feedback on the articles they are writing on the treatment of animals. But it doesn't always work this way. Animal activism destroyed Nancy's marriage. Her husband of ten years was a military man who was hostile to her increasing dedication to animal liberation; he wanted her to play the role of good army wife.

"So eventually," she told me, "I had to make a choice." She chose the animals.

Fran and her husband were on a similar collision course.

"How does your commitment to animals affect your relationships with other people?" I asked her.

She sighed. "My husband and I have lots of fights about it. He is a meat eater and thinks that people who wear fur are no worse than people who eat meat. Over the years, it has gotten worse. Now he throws my mail away because I send so much money to animal organizations."

I put the odds that they are still together at zero.

Lifestyle conflicts fell particularly hard on single activists who were looking for like-minded dating partners. An attractive twenty-something named Elizabeth told me, "My beliefs definitely interfere with my social life. I won't go out with anyone who is not a vegetarian. It limits my pool of possible men. Early on, most of the men I dated were not vegetarian. I will never do that again. Having that kind of moral blockade between you and someone you are involved with is just impossible."

There are other problems that come with being a moral crusader. Sometimes the burden just gets too heavy. I asked Lucy, a special education teacher, if people think she's crazy because of the way she lives her life.

"No," she said. "Most people don't feel that I'm nuts. But sometimes *I* think I'm nuts. I drive myself crazy about it. It dominates my life. Sometimes I think I can't take it anymore. So I say to myself, I'm going to back off a bit; I'm going to loosen the rope a little. I'm going to let myself not be Jesus for a minute and be a normal human being."

Adding to their psychological burden is the fact that animal activists are constantly bombarded with reminders of animal cruelty—the meat counter at the grocery store, the smell of grilled flesh when they walk past a Burger King, a woman wearing a fur coat at the airport, the continual barrage of fund-raising emails from animal protection organizations that flood their inbox: *Your donation will help us stop the baby seal hunt! Let's put an end to puppy mills! Shut down factory farms!*

Sometimes moral commitment can become overwhelming. Susan had insomnia because her dreams were haunted with images of animal mistreatment. Maureen and her husband were forced to declare bankruptcy because they donated all their money to animal rights organizations. And Hans, a sixty-two-year-old German-born businessman, is suffering from compassion fatigue. "I have come to near emotional collapse," he tells me.

"I am burning out. My life is so full of animal rights now that I have no time anymore. I have thrived on this in the past, but this year it came to the point where I said, 'I can't do it anymore. I just don't have the strength.'"

Like most individuals who take moral issues to heart, animal rights activists march to the beat of a different drummer. But the vast majority of activists are not fanatics. Most of the activists I have met over the years have been intelligent, articulate, friendly, and completely sane. Nonetheless, it can be hard to have a meaningful conversation with true believers, impossible to find a middle ground. Good luck explaining to Lucy why you think some animal research might be justified. I asked her if she ever had moments of doubt, if she ever thought that maybe there are circumstances in which it might be permissible to use a pig heart valve to save a human life.

"No." she said. "I definitely have the sense that what I am doing is right. And if you argue with me I am not going to listen. Because I *know* I am right."

That's kind of a conservation stopper.

A NEW KIND OF TERRORISM

Blowing off your adversaries is one thing; blowing them up is quite another. At 4 AM on March 7, 2009, UCLA neuroscientist David Jentsch awoke to the blare of a car alarm. He looked out his bedroom window and saw his Volvo in flames. He ran outside and grabbed a garden hose. Jentsch lives in one of those L.A. neighborhoods that are prone to runaway wildfires. The branches of the tree above his car were already burning when the fire department showed up. The whole neighborhood could have been taken out if the firefighters had gotten caught in traffic.

Two days later, a group called the Animal Liberation Brigade released a communiqué that said, "David, here's a message just for you. We will come for you when you least expect it and do a lot more damage than to your property. Wherever you go and whatever you do, we'll be watching you as long as you continue to do your disgusting experiments on monkeys."

Jentsch was not completely surprised that he had been targeted. In recent years, nearly a dozen UCLA researchers have been subjected to animal rights terrorist attacks. (The animal liberation underground describes these attacks as "direct action." They reserve the term "terrorist" for the people they don't agree with—fur farmers, slaughterhouse owners, animal researchers.) While most victims of these incidents decide to lay low, David Jentsch fought back. He formed an organization called UCLA Pro-Test to defend animal research on campus, and the group staged a rally in support of animal experimentation. This has not made him friends among the animal rights fringe. He regularly gets abusive emails: *David Jentsch, I want all your children to die of cancer, and I want you to watch them die. I hope you die a horrible death too.*

Social scientists have found that most terrorists are moral absolutists who are motivated by a combination of idealism, anger, religious zeal, and the human penchant to place the blame for injustice on villains. Gerard Saucier of the University of Oregon and his colleagues analyzed the thinking of a dozen kinds of militant extremists. The common elements to these disparate groups include: the belief that peaceful tactics don't work; the belief that the ends justify the means; the belief that utopia is around the corner; a belief in the need to annihilate evil; the demonization of the opposition; and the framing of conflicting moral visions as war.

You see all of these in the tiny, violent wing of the animal liberation movement—the arsonists and bomb throwers, the spray-painters, the lab animal "liberators." You also hear it in the words of individuals like Jerry Vlasak, a physician who is a press officer of the North American Animal Liberation Front. In 2004, Vlasak told an Australian television reporter, "Would I advocate taking five guilty vivisectors' lives to save hundreds of millions of innocent animal lives? Yes, I would."

Between 1993 and 2009, antiabortion extremists in the United States murdered eight people and nearly killed seventeen others. No one has died in an animal rights attack, but it is probably just a matter of time. In 2002,

James Jarboe, head of the Domestic Terrorism Section of the FBI, testified before Congress that the animal rights and environmental extremists were among the most serious domestic terrorism threats. (Animal activists argue that under the Bush administration, the FBI downplayed right-wing terrorist activities such as attacks on abortion clinics, and amped up the rhetoric on the animal rights and environmental movements.) According to the FBI, militant animal rights and environmental groups have inflicted $110 million worth of damage to animal facilities. In 2009, they had 170 extremist incidents under investigation. In 2006, Congress enacted the Animal Enterprise Terrorism Act, which increased the penalties for economic or personal harm inflicted as a result of illegal animal rights activities. And on April 4, 2009, an animal activist named Daniel San Diego was added to the FBI's Most Wanted Terrorist list. He is alleged to have bombed two corporate offices associated with animal testing.

Why have biomedical researchers and not hunters or slaughterhouse owners become the main target of the terrorist wing of the animal liberation movement? After all, the numbers of animals used in research are minuscule compared to the 10 billion animals killed in slaughterhouses or the untold millions of wild animals killed or wounded by hunters each year. The overwhelming majority of animals used in research are rats and mice, creatures most people would not hesitate to personally kill on sight (or at least pay someone else to). And animal research is certainly more defensible than eating creatures because they are tasty or shooting them for sport.

The North American Animal Liberation Front acts as a support group for the violent faction of the animal rights movement. It issues press releases announcing the latest activities of the razor-blade mailers, firebombers, and car vandals. They offer insight into the types of research that make the radicals go ballistic. Based on ALF press releases, I offer three pieces of advice for young researchers who don't want razor blades coated with rat poison showing up in their mailboxes: *Avoid California, primates, and the brain.*

Nearly 75% of ALF press releases described attacks on researchers in the Golden State. These statistics may be biased by the fact that ALF spokesman Vlasak lives near Los Angeles. However, the Foundation for Biomedical Research, a group that monitors animal rights attacks nationwide, reports that more than twice as many animal rights attacks occur in California as in any other state.

Besides living in California, the biggest predictor of whether a researcher will be targeted by animal activists is the species they work with. Most of David Jentsch's experiments involve mice and rats. But what drew the arsonists to his house was that he occasionally uses vervet monkeys in some studies. Researchers who study chickens, lizards, gypsy moths, tobacco worms, trout, spiders, parrots, mice, and rats are rarely bothered by animal activists. Rather, the targets are usually scientists who work with either monkeys or pet species (usually cats). In terms of the amount of suffering associated with animal research, this does not make any sense. At UCLA, for example, about 75,000 mice are used in research every year compared to several dozen monkeys. Three-fourths of the attacks posted on the ALF Web page were directed at primate researchers, even though monkeys and apes make up fewer than 1% of animals used in research. In contrast, only 9% of attacks described in the communiqués were directed at scientists experimenting on rats and mice—the animals used in the vast majority of biomedical studies.

Like most University of California researchers who have been attacked by anti-vivisectionists in recent years, Jentsch is a neurobiologist. His research concerns the neural mechanisms underlying schizophrenia and the effects that drugs like angel dust, ecstasy, cocaine, and nicotine have on brain cells. Why do animal liberation terrorists focus more attention on scientists seeking treatments for mental illness, drug addiction, and blindness rather than researchers working on diseases like cancer or viruses like HIV? The reason is these studies often involve primates because their brains are more similar to ours than are rodent brains. The ultimate goal of violence-prone animal liberationists is the elimination of all animal research. (A fruit-fly geneticist at the UC Santa Cruz was targeted.) However, activists in the shadow world have made a strategic decision to concentrate on researchers who study a handful of species the average person will empathize with. Pictures of cute monkeys make for better fund-raising brochures than photographs of beady-eyed albino rats.

Given the toll that the threats and emails and the car bombing have had on his life, David Jentsch has a surprisingly positive attitude toward animal rights activists. Since the attack, Jentsch has met with local animal protectionists to try to develop what he calls a talking relationship with them, to at least try to get a conversation going. Most of the activists he has encountered are reasonable people who are opposed to the harassment of individual researchers. They do not appreciate the attention the media gives to Animal Liberation Front. Jentsch thinks that only a tiny group of zealots have been responsible for all the death threats, obscene emails, and firebombings. This view is supported by a Department of Homeland Security report that put "hard-core" ALF membership at about 100 individuals in 2001.

"But it only takes one person to plant a bomb," David Jentsch reminds me.

MORAL CONSISTENCY AND ANIMAL LIBERATION PHILOSOPHY

Putting an incendiary device on someone's front porch is animal liberation run amok. But most animal activists are not violent fanatics. To understand the larger questions that direct action on behalf of animals raises about human moral thinking, it helps to look at the ethical theories on which the modern animal liberation movement is based.

Much like journalism, ethics boils down to who, what, and why questions: who is entitled to moral concern, what obligations we have to them, and why one course of action is better than another. The technical literature concerning our obligations toward other species is vast, complicated, and, for the most part, boring. The philosophical case for giving moral status to animals has been made by Aristotelians, feminists, Darwinians, Christian right-wingers, and postmodern leftists. However, the major intellectual paths to animal liberation lie in the two classic approaches to ethics, utilitarianism and deontology. Utilitarians believe that the morality of an act depends on its consequences. Deontologists, on the other hand, argue that the rightness or wrongness of an act is independent of its consequences. They believe that ethics are based on universal principles and obligations (the term *deontology* comes from the Greek word for obligation, *deon*). In other words, you should keep your promises not because bad things will happen if you break them, but because you made them.

The application of utilitarian principles to the treatment of animals was first made by the eighteenth-century philosopher Jeremy Bentham, who argued that acts should be judged on the degree that they increase pleasure and decrease pain. His twist was to insist that other species be included in the calculus. He wrote, "The question is not 'Can they reason,' nor 'Can they talk,' but 'Can they suffer?'" Peter Singer of Princeton University—arguably the most influential living philosopher—used this line of thinking as the cornerstone of his 1975 book *Animal Liberation*, which jump-started the contemporary animal liberation movement.

Ironically, while the book is often called the bible of the animal rights movement, Singer's argument for animal liberation is not based on the idea that animals (or humans, for that matter) have inherent rights. Rather, his case is based on simple fairness. Singer lays his position out in a single sentence: "The core of this book is the claim that to discriminate against a being solely on account of their species is a form of prejudice, immoral and indefensible in the same way that discrimination on the basis of race is immoral and indefensible." He refers to bias toward the interests of your species and against members of other species as "speciesism," which he feels is as morally repugnant as racism and sexism. Singer's argument rests on the notions that all sentient creatures (organisms capable of experiencing pleasure and pain) have the same stake in their own existence and that suffering is the ultimate moral leveler. "From an ethical point of view," he writes, "we all stand on an equal footing—whether we stand on two feet, or four, or none at all."

The intellectual architect of the other path to animal liberation—the deontological route—is Tom Regan, who wrote the important 1983 book *The Case for Animal Rights*. Regan starts with the idea that humans and some animals deserve moral consideration because they are the "subjects of a life" and therefore have inherent value. By this, he means that they possess memories, beliefs, desires, emotions, a sense of the future, and a sense of identity of their own existence over time. Regan believes that if a creature has inherent value, it is wrong to treat it as a mere thing to be used or discarded. I am often irritated with the balky computer on my office desk. Regan would say that it would be morally permissible (and I would add, psychologically rewarding) to heave it out my third floor window. But, according to Regan, it would *not* be ethical for me to toss an irritating student out the window because he or she wants to argue with me over a test grade. Nor would he approve of throwing my cat out the window because I tire of her nagging me to rub her belly while I am trying to write an article about animal ethics. Regan's reasoning is that, as subjects of a life, both the argumentative student and Tilly have fundamental rights that my computer does not possess. Most important, they have them in equal measure. Among these are the right to be treated with respect and the right not to be harmed.

Regan and Singer differ on some issues. For example, they disagree as to why it would be wrong for me to toss a student and/or a cat out a window. Singer would say it is because they would suffer, not because they have inherent rights. And while Singer is not, at least in principle, opposed to painlessly taking a human or a nonhuman life in some circumstances, Regan is. The two philosophers are, however, in accord on most of the big issues. They both acknowledge that there are important differences between humans and other animals, but they believe that these are not relevant to whether a creature deserves moral consideration. The logical extension of both Regan's rights argument and Singer's utilitarian logic is that we should not eat animals, hunt them, or otherwise avoidably cause them to suffer. Practices such as factory farming, using animals for research, caging them in zoos, or trapping them for their fur are immoral under both positions.

CAUGHT IN THE GRIP OF A THEORY:
ANIMAL ETHICS AND THE LIMITS OF LOGIC

Inevitably, ethics involves drawing lines. Singer originally drew the line "somewhere between the shrimp and the oyster" while Regan set the bar at the level of mammals and birds at least one year old. (He bends the rules a bit by saying that basic rights should also extend to human infants.) Singer and Regan both recognize that humans live in the real world, not in an imaginary moral ether. Thus they are willing to make the occasional compromise to accommodate common sense. Both, for example, implicitly recognize that some species warrant more concern than others. Singer has invested more energy into promoting a campaign to gain legal standing for great apes than he has to banning mousetraps. And Regan says that if four normal humans and a golden retriever are in an overloaded lifeboat that can only carry four, the dog goes overboard. He writes, "Death for the dog is not comparable to the harm that death could be for any of the humans."

But what happens when you refuse to draw moral lines in the sand? Ralph Waldo Emerson famously wrote, "A foolish consistency is the hobgoblin of little minds." In animal ethics, foolish consistencies are exemplified by Joan Dunayer, author of the book *Speciesism*. By insisting on a combination of animal rights literalism and uncompromising adherence to moral consistency, she constructs a set of impossible ethical standards. She also illustrates what happens when you take logic too far.

Dunayer would, of course, consider me a speciesist because I eat meat. More surprising are her vicious attacks on the intellectual elite of the animal liberation movement and on radical animal rights groups that do not measure up to her standards of ideological purity. Dunayer, for example, adamantly opposes any effort to reduce animal suffering by replacing a cruel practice with a less painful alternative. She denounces PETA because they pressured the fast-food industry to make life better for chickens on factory farms. Bigger cages are unacceptable; for Dunayer it is empty cages or nothing. She is mad at Tom Regan for arguing that the dog should be booted out of the hypothetical life raft ahead of any human.

Peter Singer merits a special place on her list of animal rights cop-outs. She disapproves of Singer's efforts to single out chimpanzees and gorillas as candidates for legal standing. (I assume she would also disapprove of Singer's statement that he would not have much compunction about swatting a cockroach; insects don't suffer much, he says.) Dunayer is also unhappy that Singer believes that a human life counts more than the life of a chicken. He argues, for example, that the deaths of 3,000 humans on September 11, 2001, were a much greater tragedy than the deaths of the 38 million chickens killed in American slaughterhouses that day. Dunayer disagrees. In her eyes, chickens deserve even more moral consideration than humans. She writes, "Singer's disrespect for chickens is inconsistent with his espoused philosophy which values benign individuals more than those who, on balance, cause harm. By that measure, chickens are worthier than most humans, who needlessly cause much suffering and death (for example, by wearing animal-derived products)."

Philosophers have a phrase for what happens when people take logic to bizarre extremes. They say you are "caught in the grip of a theory." Dunayer falls into the grip by making two assumptions that seem reasonable until you play them out. The first is that all creatures who can experience pleasure and pain should be treated equally. The second is that all it takes to experience pain is the simplest of nervous systems.

In Dunayer's own words, here are some of the logical consequences of these apparently innocuous assumptions, taken from *Speciesism*:

- "Because all sentient beings are equal, we're perfectly entitled to save the dog over any of the human beings." (p. 97)
- "Wasps need a legal right to life." (p. 141)
- "Our moral obligations need to include insects and all other beings with a nervous system. . . . these animals include comb jellies, cnidarians such as jellyfishes, hydras, sea anemones, and corals." (p. 127)

Joan Dunayer lives in a moral universe that should cause even hardcore animal activists to shudder. Can a reasonable person really believe, as Dunayer apparently does, that one should flip a coin when deciding whether to snatch a puppy or a child from a burning building, or that duck hunters should be imprisoned for life?

The problem for animal liberationists is that Dunayer is right. If you take the charge of speciesism literally, if you refuse to draw any moral lines between species, if you really believe that how we treat creatures should not depend on the size of their brains or the number of their legs, you wind up in a world in which, as Dunayer suggests, termites have the right to eat your house.

HOW SHOULD A GOOD PERSON ACT?

I hate my Inner Lawyer. He usually pops up when I am taking a shower in the morning, or driving mindlessly on some mountain road with the radio turned off. He's my Jiminy Cricket, my Obi-Wan Kenobi. He asks me inconvenient questions. The neuroscientist Joshua Greene says he lives in a little section of my brain behind my eyebrows called the DLPFC—the dorsolateral prefrontal cortex. Using a brain-imaging technique called functional MRI, Greene has found that our DLPFC lights up when we try to think logically about tough moral issues. My Inner Lawyer showed up a couple of days ago when I was hiking in the Smokies.

I.L.:   Hal, it's me—your Inner Lawyer.

HAL: Go away.

I.L.:   Just listen for a second. Pretend that it is 1939 and you are living in the quaint little village of Dachau outside of Munich. Every day you see the smoke from the chimneys behind the fence of the new "camp" and know that Hitler's goons are working the ovens overtime to exterminate Jews, Gypsies, and homosexuals. Rumor has it that Nazi doctors are even conducting painful medical experiments on some of the prisoners. Your friend Heinz asks you to help him plant a bomb under the SS guards' barracks. "By killing them we will save lives," Heinz whispers. "We will send a message to the world." Hal, do you think you should help Heinz blow up the barracks and perhaps save thousands of people?

HAL: I wouldn't have the guts.

I.L.: I know. But just pretend you are both a brave and a good person.

HAL: *sigh* . . . OK, a good person with courage would be justified in taking direct action to help prevent genocide, even if it meant killing barracks full of Nazis.

I.L.: I agree. Now assume you have read books by the intellectuals of the animal rights movement. They have convinced you that speciesism is the moral equivalent of racism, that the suffering of a monkey in a laboratory is morally no different than the suffering of a human child. You also know that 60,000 monkeys a year are used in research in the United States. And, you agree with Jerry Vlasak of the Animal Liberation Front that killing just one or two primate researchers might put all these studies to a halt.

Now, answer this question. Would a good and brave person like you be justified in firebombing the home of a scientist who was addicting monkeys to cocaine in order to study the brain chemistry of addiction?

HAL: No, it would be against the law.

I.L.: But murder was against the law in Nazi Germany, and you said that was OK.

HAL: That was different.

I.L.: Why?

HAL: Because keeping monkeys in a lab and even killing them for science is not the same as killing Jews in a concentration camp.

I.L.: But if you *believe* there is no morally relevant difference between a person and a monkey, wouldn't you be justified in harming primate researchers?

> *Hmm . . . I am thinking that I.L. has me by the balls. But*
> *then I am saved by a dim recollection of a lecture I heard years*
> *ago on the ethical theory of Immanuel Kant.*

HAL: Gotcha. Kant argued that that you should always act in the way that you would want everyone to act in the same situation. I would not want to live in a world where every whacked-out moral crusader with a gun would be allowed to shoot people he thought were doing harm to his pet cause—old-growth forests, fetuses, or frogs. The world would be chaos.

I.L.:   But Hal, you did say it would be justified to blow up the Nazi concentration camp guards. How do you know when it is moral to take an illegal action and when it is not?

HAL: You just KNOW. It's common sense!

I.L.:   So, you think that when it comes to killing someone, you should rely on your own common sense, your personal moral intuition? Would Kant go for that?

HAL: Get out of my life, asshole.

IN MATTERS OF MORALS, YOU CAN'T TRUST YOUR HEAD. . . . OR YOUR HEART

My Inner Lawyer raises a question at the center of all human morality, not just animal ethics: How do we know what is right? There are two places where we can turn for moral guidance: our head and our heart. The problem is that you can't rely on either of them.

First, head. Cognitive psychologists have repeatedly demonstrated that human thinking is, in the words of the behavioral economist Dan Ariely, "predictably irrational." Researchers have identified dozens of types of bias that unconsciously warp the way we think. They have great names: the Lake Woebegone Effect, Myside Bias, the Gambler's Fallacy, the Barnum Effect, Naïve Realism. The list goes on.

Dunayer's conclusion that a spider and a human child have the same moral status is both logical and absurd. It illustrates how pure reason can lead us to completely warped ethical standards, that even when we apply the rules of logic, things can go awry when we are making ethical decisions. Rob Bass, my philosopher friend, disagrees. He believes that if you correctly apply formal deductive logic to premises that are true, you will always end up with a correct conclusion. In theory, he may be right. However, psychologists have repeatedly found that humans vary greatly in their ability to think rationally about moral issues. Further, there is abundant evidence that there is almost no relationship between the sophistication of a person's ethical thinking and how they actually behave.

Even Tom Regan and Peter Singer, both first-class intellects, get into trouble by taking moral consistency too seriously. For example, in his life-boat scenario, Regan concludes that the dog goes overboard first. Then he takes it a step further and says that you also should toss a million dogs overboard if it would save a single human. But, at the same time, Regan argues that it is wrong to sacrifice a million mice for biomedical research that might ultimately save millions of human children.

Logic also leads Peter Singer to conclusions that most people would find unnerving. In *Practical Ethics,* he shows that the logical upshot of his utilitarianism is that it might be permissible to euthanize a permanently disabled infant if the child's mother might subsequently give birth to a healthy child. And Singer has raised the possibility that sexual interactions between humans and animals are not necessarily harmful to man or beast. While his remarks were taken out of context, his comments were met with howls of protests by both the press and animal advocates.

The adherence to cold logic in moral decision making has led some philosophers to conclude that it may be preferable to use impaired human infants rather than monkeys in biomedical experiments, that arson can be a legitimate agent of social change, and that the life of an ant and that of an ape are of equal moral value. So much for relying on our heads when it comes to thinking about animals.

What about our hearts? Is moral intuition better than logic at resolving the moral conundrums in our dealings with other species?

Unfortunately, no. If anything, in matters of morality, our hearts are even more prone to error than our heads. Intuition (and its handmaiden, common sense) are subjected to the whims of a host of morally irrelevant factors—how big an animal's eyes are, its size, whether it was the mascot of your high school football team—and the evolutionary history of our species. Moral intuition told my friend Sammy Hensley that his hounds did not mind spending their lives chained to a doghouse and tells Japanese fisherman that there is nothing wrong with slaughtering dolphins because they are "fish." My moral intuition tells me that it is OK to eat meat (particularly if it is labeled "cruelty-free") but my friend Al's moral intuition tells him that meat is murder. For thousands of years, it was common sense that slaves were property and that homosexuality was a crime against nature. And moral intuition told the 9/11 airplane hijackers and the arsonists who planted the firebomb under David Jentsch's car that they had the moral high ground.

WHO HAS THE MORAL HIGH GROUND

I am confused and need another opinion on the "ought to"s of ethics and animals, so I send an email to Gayle Dean, an animal advocate who takes ethics seriously.

> Gayle, from an animal liberation perspective, is there any difference between the 9/11 terrorists and "direct action" ALF types who attack scientists? After all, both groups were completely convinced that they have the moral high ground.

She writes back.

> There is a big difference between groups that are *convinced* they have the moral high ground, and the ones that actually *have* the moral high ground. The difference lies in the truth of the matter. During slavery, many people risked their lives to illegally help slaves because they were convinced they had the moral high ground. In fact, they actually did have

the moral high ground. The same for those who helped the Jews escape
from the Nazis.

It is after midnight. Tilly has fallen asleep on the rocking chair in my
office. I am tired. I write back.

Gayle, I agree with you about Nazis and slavery. But here is my question—
how can we be sure what the moral truth is in these matters? Doesn't it
boil down to personal opinion, your own moral intuition?

The next morning she responds:

I agree that is difficult to know the truth of the matter. But moral truth
certainly does NOT boil down to personal opinion!

I'd like to think she is right, but the longer I study human-animal interactions, the more I have my doubts.

The moral psychologist Jonathan Haidt says, when push comes to
shove, we are all hypocrites. After twenty years of studying how people
think about animals, I have come to believe he is right. You do, of course,
run into the occasional exceptions. Lisa, for example, is a vegan who does
not take antibiotics or let her cat outdoors where it might have fun stalking
birds. But the vast majority of us are inconsistent, often wildly so, in our
attitudes and behavior toward other species. What are we to make of this?

In the 1950s, the social psychologist Leon Festinger proposed one of
the most influential theories in psychology—that when our beliefs, behavior, and attitudes are at odds, we experience a state that he called cognitive dissonance. Because dissonance is uncomfortable, people should be
motivated to reduce these psychic conflicts caused by inconsistency. We
might, for example, change our beliefs or our behaviors, or we might dis-
tort or deny the evidence.

The environmental philosopher Chris Diehm (who is a vegan) is opti-
mistic. He says that when he discusses with people inconsistencies in how
they treat animals, they often make an effort to change, or at least they try

to justify their behavior. He writes, "We recognize that our relationships to animals take widely disparate, apparently contradictory paths: We have cats in our houses but cows on our plates. When people have this inconsistency pointed out, they try to make sense of it, or remove it to the point where they are comfortable with it. The drive for consistency seems to be a good thing, and exposing inconsistencies is a deep motivator for moral reflection and development."

Chris is a philosopher. He is impressed by the need of humans to achieve logical coherence in their beliefs and behaviors. I am a psychologist. I am more impressed by our ability to ignore even the most blatant examples of moral inconsistency in how we think and behave toward animals. In my experience, most people—be they cockfighters, animal researchers, or pet owners—remain stubbornly oblivious (other than an occasional uncomfortable laugh) when you point out the paradoxes and inconsistencies in our personal and cultural treatment of animals.

So moral consistency is elusive, if not impossible, in the real world, and both head and heart can lead us astray in how we think about the treatment of animals. Perhaps, as shown in the next chapter, we should look to the lives of virtuous individuals rather than abstract philosophical treatises for guidance in our lives with other species.

10

The Carnivorous Yahoo
Within Ourselves

DEALING WITH MORAL INCONSISTENCY

> The fact that you can only do a little is no excuse for doing nothing.
>
> —JOHN LE CARRÉ

The central character of the Nobel Prize–winning author J. M. Coetzee's novel *Elizabeth Costello* is a visiting scholar who delivers a series of lectures on the moral status of animals at a major university. After one of her public talks, an audience member raises her hand: "Are you not expecting too much of humankind when you ask us to live without species exploitation, without cruelty? Is it not more human to accept your own humanity—even if it means embracing the carnivorous yahoo within ourselves?"

Good question. Coping with our inner yahoos is a central theme of ethics, psychology, and religion. The yahoo goes by different names. Freud calls it the Id. George Lucas called it Darth Vader. When Jesus warned that the spirit is willing but the flesh is weak, he was warning about the yahoo. George Jones sang about it in "Almost Persuaded." Evolutionary psychologists trace its origins to the Pleistocene, and neuroscientists say it divides its time between your frontal lobes and your limbic system.

Jonathan Haidt, the psychologist who has explored the moral ramifications of the yahoo more than anyone else, compares it to an emotional elephant being ridden by a rational rider. The elephant is large and usually calls the shots, albeit unconsciously. The rider is weaker than the elephant, but smarter. With practice, the rider can exert some control over the elephant. I have argued in this book that the paradoxes that characterize our relationships with other species are the unavoidable result of the perennial tug of war between the rational part of us and the yahoo within. But what are the implications of living in a world that is morally convoluted, in which consistency is elusive, and often impossible? Do we throw up our hands in despair? Does moral complexity mean moral paralysis?

No. I have met lots of animal people who have come to terms with their carnivorous yahoo. They work for animals in different ways and on different scales. Most of them do small things that help animals and make them feel good about themselves. Some of them cut back on their meat consumption or adopt a shelter dog. Others donate money to PETA or the World Wildlife Fund or pull over to the side of the road and carry a box turtle in the middle of a highway to safety.

Others work for animals in a big way. Michael Mountain is one of them.

HELPING ANIMALS ON A GRAND SCALE: A GLOBAL KINDNESS REVOLUTION

A man walks into a bar. . .

The bar was in the Sheraton Hotel in Raleigh, North Carolina, where I was attending a conference on human-animal relationships. The man was in his early sixties, tall, wiry, reddish hair, neatly trimmed beard, outdoorsy—a ruddy Abraham Lincoln. He looked around and saw that the only empty seat in the place was next to me.

"Do you mind if I sit here?" English accent. Oxbridge.

"No. Have a seat. I'm Hal Herzog."

"Michael Mountain, Best Friends Animal Society"

"Oh, yeah—I think I've heard of that. Out in the middle of nowhere, right? In the desert."

"Yes. Kanab, Utah."

We order beers.

I asked him about Best Friends. He says that it was founded twenty-five years ago by a ragtag band of animal lovers who dreamed of a place where homeless dogs and cats would never be euthanized. He tells me that it has grown to a $35 million operation (the same size as PETA); that Best Friends rescued 6,000 animals during Hurricane Katrina; that they are no longer Best Friends Animal *Sanctuary* but have reorganized as Best Friends Animal *Society*, a nationwide network of people and grassroots community organizations, all devoted to saving animals.

I am impressed. But I am more impressed when our discussion turns to how people think about animals. He gets it: Human attitudes toward other species are inevitably paradoxical and inconsistent. He confesses some of his own moral lapses. He is a vegan and does not eat any animal products. But he purchases pigs' ears for his dogs to chew on. The dogs love them but Michael says he can't stop thinking about the poor pigs.

Then he tells me about the convoluted ethical guidelines that he follows in dealing with the biting horseflies that buzz around his house in the summer.

"Here is my rule," he says. "If I am walking outside and a horsefly bites me, it is permissible for me to swat it, just like you would a mosquito. However, if the horsefly comes into my house, I have to rescue it and take it outside." Smiling, he adds, "Where it will bite me the next time I go for a walk."

"Huh? That's completely ass-backwards," I say. "It should be OK to kill a fly that has invaded your house, your home territory, and not OK for you to kill one outdoors when you are on its home territory. Is there a rationale for your rule?"

He laughs.

"Of course. There is always a rationale. But a rationale is not necessarily rational. I suppose my rule is much like the philosophy behind Best Friends. You can't save all the animals in the world, but the ones that come into your care, you are responsible for. So, once the fly enters my house, I have a responsibility to treat it with kindness."

Here is a rare bird: a morally serious person who can laugh at himself.

Having recently stepped down as Best Friends' president, chief fund-raiser, and the editor of its magazine, Michael Mountain has teamed up with a young entrepreneur named Landon Pollack on a couple of new projects. One of these is rehabilitating the image of pit bulls in the United States. The other is organizing a global community of people who care about animals and the natural world. Its name is Zoe, the Greek word for life.

"We want to lead a global kindness revolution that will transform the way people relate to animals, nature, and each other."

A global kindness revolution? It sounds grandiose, possibly insane. But the guy seems like the real deal.

I glance at my watch. We have been talking for two hours. It's closing in on midnight and we are the only ones left in the bar. As we get up to leave, Michael says that I should come out to Kanab, to talk more and see what Best Friends is about.

"Maybe I'll take you up on it."

I get back to my hotel room and telephone Mary Jean. "Would you be interested in flying to Utah next summer?"

THE ANIMAL SANCTUARY IN THE MIDDLE OF NOWHERE

Mary Jean and I get off the plane in Las Vegas, pick up a black Hyundai from Avis, and head north up I-15. To get to Kanab, Utah, you drive two hours to Saint George, turn off the interstate, and drive another couple of hours on a two-lane blacktop, dipping into Arizona through Colorado City (usually referred to in the media as a "polygamy enclave") and the Kaibab Paiute Indian Reservation, on into Kanab, a town with one traffic light and 3,769 full-time (human) residents.

The next morning, we head out to Best Friends, five miles out of town on Highway 89. I expect the sanctuary to look like a big petting zoo. Wrong.

Nestled next to the Delaware-sized Grand Staircase-Escalante National Monument, the 3,700-acre sanctuary is encompassed within another 30,000 acres the organization leases from the Bureau of Land Management. The vista is of biblical proportions—big sky, miles of sandstone cliffs, and mesas that remind me of the colors in the giant box of Crayola crayons I used to fight over with my sister: Brick Red, Burnt Sienna, Mahogany, Raw Umber, Carnation Pink, Copper, Maroon. I feel as if I've seen this place before. Later I learn that I have. Many of my favorite TV shows as a child were shot in or around Angel Canyon—*The Lone Ranger, Rin Tin Tin, Lassie, Have Gun Will Travel, Gunsmoke*—even *Death Valley Days,* starring Ronald Reagan. Since the 1920s, nearly 100 feature films have been shot here, including *Planet of the Apes, The Greatest Story Ever Told, The Man Who Loved Cat Dancing,* and *The Outlaw Josey Wales.*

Best Friends offers daily tours for the 30,000 visitors that stop by every year, but Michael has arranged a special behind-the-scenes tour for us. Our guide is Faith Maloney, a cheery sixty-five-year-old Englishwoman who seems to know the name of every one of the 1,700 rescued dogs, cats, pigs, horses, rabbits, donkeys, peacocks, guinea pigs, and parrots in residence. Michael and Faith were part of a small group referred to reverently as "the founders." Best Friends grew from the vision of some young idealists in the mid-1960s who had a desire to do some good in the world. ("We were not hippies," Michael warns me. "If anything, we were the anti-hippies. For example, our rule was no drugs.") After a stint in the Yucatan, the group became involved in politics, religion, and social services before disbanding, but some of them reunited and discovered that they had a common interest in saving animals.

In the early '80s, the group stumbled on Kanab Canyon, which they renamed Angel Canyon. Despite its remoteness, they decided that the canyon was just the place to establish a home for animals no one wanted. They did not foresee that their small shelter in southwestern Utah would become one of the nation's largest animal protection organizations; that an army of volunteers would one day walk dogs, hose down pot-bellied pigs, and shovel horse manure; that the sanctuary would be the subject of a hit television series (National Geographic Channel's *Dogtown*); or that Best Friends would play a pivotal role in the most successful campaign in the history of the American animal protection movement—the establishment of community-based spay/neuter and adoption programs that has reduced the number of dogs and cats killed each year in animal shelters from 17 million to 4 million.

We meet Faith at the visitor center and walk over to Piggy Paradise (Best Friends has a penchant for cute names). We see a volunteer from Virginia exercising a pot-bellied pig by tempting it with pieces of no-fat popcorn. We talk to a woman farrier who is rebuilding the shattered hoof of an enormous draft horse that had been abandoned at a local dump. Then we jump into Faith's car and take off to Casa del Calmar. It is a house—a real house (no animals are kept in cages at Best Friends)—for cats with incurable conditions like feline leukemia. The place is spotless, nary a whiff of pee, even though purring cats are draped everywhere. Faith explains that the focus of Best Friends is on animals with special needs—a three-legged cat, a dog with a lump on its throat the size of a baseball, an eagle with a broken wing. Most of them come from other shelters that can't give them the longer-term care they need. This is the shelter of last resort for the blind, the deaf, the psychologically damaged.

After brief stops at Bunny House, Horse Haven, Parrot Garden, and Wild Friends (a rehab center for turtles, owls, hawks, bob cats, and songbirds), we turn into Dogtown Heights, a ninety-acre complex that is home to some 400 dogs. At Old Friends, a facility for aging dogs, I meet Ruby Benjamin, an energetic seventy-eight-year-old psychotherapist from Manhattan who volunteers for a couple of weeks at Best Friends every year. "My heart is here," she tells me. "When I come to Best Friends, it is like getting a big hug."

In the main building at Dogtown, we are introduced to Cherry, a smallish black and white pit bull lying placidly on a cushion under the desk of a young woman who is typing on a computer. The doleful-eyed Cherry looks completely laid-back—you would never suspect that she was one of nearly two dozen of quarterback Michael Vick's fighting dogs that were taken to Best Friends in the aftermath of the raid on Bad Newz Kennels in Virginia.

Then it's a quick tour of the clinic. A visiting intern from a California vet school is spaying a cat in the surgical theater under the supervision of one of the six staff veterinarians. We stick our heads in the hydrotherapy room. Then, a technician proudly shows us his new state-of-the-art computer imaging X-ray machine. The clinic is better equipped than some of the hospitals I've been in.

Faith has arranged for us to have lunch with Frank McMillan. Dr. Frank, as he known at Best Friends, is a veterinarian and an authority on the mental health of companion animals. He is the person in charge of rehabilitating the Vick pit bulls. He says that these animals are suffering from a canine version of post-traumatic stress disorder. Horribly mistreated at the hands of dogfighters, their primary symptom is not aggression but fear. Frank and a team of animal behaviorists have been working with them for a year and a half. He originally expected the worst, but he has been pleasantly surprised. Twenty-one of the twenty-two dogs, he says, are making real progress. Several have passed a standardized canine "good citizen" test. And, so far, two have been placed in homes and one is in foster care.

The next morning, I return to spend the day as a volunteer. I am assigned to Dogtown. I am greeted by a man named Don Bain, a retired banker from Texas. He and his wife stumbled on Best Friends, fell in love with the place, and wound up buying a house in Kanab. Now he works in Dogtown as the "puppy socialization coordinator," which is, for my money, the world's greatest job title. He assigns me to work with a staff member named Terry. My job is to help Terry feed a dozen or so dogs. One of them is Shadow, one of the Vick pit bulls. Shadow takes one look at me and snarls. Terry tells me that he is like that sometimes with strange men.

That's OK. I know that it is not his fault. He is a victim. At most animal shelters, he would have gotten the needle long ago. At Best Friends he has a home for life, even if he turns out to be unadoptable.

Then it is walk time.

Terry teams me up with another volunteer, Dora, a woman from Kansas

City who works at Home Depot. She is driving home from San Francisco and has taken a two-day detour so she could spend a day volunteering at Best Friends. Dora snaps a leash on a brown Lab-ish looking dog named Cinderella. I get Lola, a mixed-breed I instantly fall in love with because she is the long-lost twin of my childhood dog, Frisky. Off we go on the walking trail through the pinion pines, juniper, rabbit bush, and prickly pear, the White Cliffs of the Grand Staircase in the distance. The dogs are happy and so are we . . . at least until Dora and I, busily talking about dogs, become hopelessly lost. Terry finds us just as the wind kicks up, the temperature drops, thunder rolls, and the rain starts to fall.

Over dinner, Mary Jean and I tell Michael about our experiences at the sanctuary. After hanging around Kanab for a week, we are astonished at what a handful of dreamers with a vision have accomplished in the Utah desert. It is a first-class operation. The bathrooms are clean and the staff returns your phone calls. More impressively, all the animals are individuals at Best Friends. The staff does not talk about "dogs" or "cats" or "horses"; they talk about James, Minda, and Moonshine (a black-and-white guinea pig). There is an eerie tranquility about the place. Everyone is *so* nice. It's all a little spooky.

The philosopher Immanuel Kant argued that humans should not mistreat animals, but only because he felt that animal cruelty makes people more violent toward other people. The vision of the founders of Best Friends was the converse of Kant's dictum; they were convinced that being nice to animals makes us kinder to other humans.

Now Michael wants to take this philosophy to a new level through Zoe, his new project and the focus of his global kindness revolution. It is still in the early stages, but Michael has put together a management team and an advisory board that consists of corporate heavy-hitters and experts from the worlds of animal protection, the sciences and humanities, communications, marketing, publishing, and social networking. They are thinking big—a series of books, perhaps a magazine and television network, a Huffington Post–type Web site where you can go for a daily hit of animal and environmental news. Zoe will be a lifestyle brand, a big tent for all kinds of people who care about animals and nature—recyclers, tree huggers, vegans and "relaxed vegetarians," people who drink free-trade coffee and people who eat cruelty-free chickens. In short, people who want to make the world a better place, people who want to reconnect with animals and nature, but aren't quite sure how to do it.

The scope of his vision tires me out. Michael thinks differently than I do. He thinks BIG. Too big for me.

I change the subject. "Have you always been good with animals?"

He surprises me. "Well, I don't get all gooey over them. I am better at organizing and editing and putting people together than I am with taking care of animals."

But then he tells me about the ants in his kitchen. "They are cute, the ants. They are basically like a sanitation department. The ants come in to the kitchen on these patrols. If they don't find anything, they go back outside. But if Miss Popsicle, my cat, has left a little piece of food some-where, they mount a military operation to transport it outside. It is really quite amazing. It takes a lot of ants to carry a piece of cat food. When I see them putting this much work into it, I try to help them out by getting them and the piece of cat food on a piece of tissue paper and take them outside."

I try to imagine a man who can chat up the CEO of a Fortune 500 corporation, who is on a first-name basis with Hollywood's A-list, but who claims he is not particularly good with animals, down on hands and knees on his kitchen floor gently brushing tiny ants onto a Kleenex and trans-porting them back to their nest, just to make their day a little easier.

THE LADY AND THE SEA TURTLES

Michael Mountain is a dreamer trying to pull off a global revolution. But most animal people are more like Judy Muzee. She owns Beach Combers Hair and Nail Studio, a beauty shop on Edisto Island, South Carolina. She spends her spare time trying to save endangered sea turtles.

The path to becoming an animal rescuer usually begins with concern for a single animal—an emaciated stray dog, perhaps, or a cat on the eu-thanasia list at an animal shelter with a three-day-and-out policy. Sea turtle people are different. They don't get much back in the way of the warm and fuzzies. Indeed, many of them will never even see one of the animals they are trying to save. It is enough for them to know when they patrol the beach at dawn that just beyond the surf, a 300-pound female loggerhead might be waiting for the sun to set so she can lumber ashore, laboriously dig a nest two feet deep in the sand, and lay a hundred squishy eggs the size of Ping-Pong balls which will hatch a couple of months later, and that, with luck, one in a thousand hatchlings will survive to repeat the process twenty-five years later.

For a beach town, Edisto is backwater. It has no motel, no putt-putt, no McDonald's, and no water park. There is, however, a grungy Piggly Wiggly supermarket and Whaley's, a slightly seedy bar with a pool table, good food, and an easy ambience two blocks off the main road near the town's water tower. Mary Jean and I were in Whaley's one Sunday evening sipping beer while waiting for our shrimp sandwiches. Mary Jean was talking to the woman sitting on the barstool next to her. My attention was divided between the NASCAR race on the television above the bar and two guys across from me talking fishing and drinking oyster shooters for dinner. (You drop a raw oyster in the bottom of a shot glass, add an ounce of Smirnoff, a little Tabasco sauce, and a squeeze of lemon. Toss it down in one gulp, and chase it with a swig of Bud Lite.)

I had watched them work their way through a dozen oysters when I heard Mary Jean say to her new friend, "Oh, you need to talk to my husband—he studies people who love animals." Judy, it turns out, was nuts over loggerhead turtles—the giant reptiles that nest on beaches from Texas to North Carolina. Mary Jean and I switched seats.

Judy told me she had moved to Edisto ten years ago from Wyoming after her marriage broke up. She worked two jobs for a couple of years and saved enough money to open her hair salon. I asked her about sea turtles, and she lit up, whipped out her cell phone, and started showing me pictures of turtle tracks on the beach, opened nest cavities, and newly hatched button-cute babies. Judy is part of a team of volunteers that patrols the beach at dawn, recording crawls (the three-foot-wide trails in the sand that nesting females leave on the beach) and nest locations, and moving vulnerable nests to safer ground. Once the eggs begin to hatch in a couple of months, she returns and records their fate—the numbers of successfully hatched eggs and dead babies.

Judy invited me to join her the next day on the dawn patrol but we were leaving town. I promised to come back.

A year later, I am sitting in Judy's living room drinking sweet tea. It is dark and cool, which is good because it is nearly 100 degrees outside with 98% humidity. She introduces me to her arthritic chocolate Lab (named OB because he only has One Ball) and to Megan, her eighteen-year-old granddaughter who also helps with the turtle rescue project. They fill me in on the basics of loggerhead reproductive biology. Female turtles spend their entire lives roaming the oceans, only coming on land every two or three years to lay their eggs. (The males never come ashore.) The nests are architectural marvels. They are shaped like two-foot-deep chemistry flasks with the egg cavity mushrooming out of the bottom of a narrow tunnel. The female excavates the nest with her rear flippers. Once the eggs are deposited, she refills the nest hole and obscures its location from predators.

Many eggs never hatch. A raccoon will dig up the nest or a ghost crab will hide out in the egg cavity, gorging itself on yolks and embryos. In about fifty days, the nestlings will pip their shells, spend a couple of days resting in the nest cavity and absorbing egg yolk, and then start digging their way to the surface. They almost always emerge at night, and instinctively know to make their way toward the surf, drawn by the open sky over the ocean and the reflection of moonlight on the water.

Loggerheads are endangered. Even under the best of circumstances, only one-tenth of one percent of the hatchlings will reach reproductive age and return back to these beaches to continue the cycle. Everything eats baby turtles. But sea turtles are not endangered because of raccoons or ghost crabs or sharks or birds. No, they may go the way of the passenger pigeon because they swallow plastic shopping bags they mistake for jellyfish, or get caught in commercial fishing nets or are poisoned by toxic chemicals or oil spills. There is also the loss of nesting habitat because of

beachfront development. At Edisto, light pollution is a big problem. Instead of heading toward the surf, hatchlings are drawn inland by the lights
from condos rented by insomniacs or from the gas station just across the
road from the beach.

The South Carolina Department of Natural Resources is in charge
of the sea turtle protection program, but it is really made possible by 800
volunteers like Judy who monitor virtually every mile of turtle beach in the
state during the breeding season. The program has several purposes. One
is scientific: The data gathered by the volunteers is invaluable. Thanks to
the efforts of the beach patrollers, biologists know the exact location of
nearly every sea turtle nest in South Carolina. They can tell you for any
given stretch of beach, the number of crawls, the number of nests, how
many eggs were laid, the proportion of dead and live hatchings, and the
sources of their mortality. In 2009, for example, 2,184 nests were located
on South Carolina beaches, 163,334 eggs hatched, and 10,503 were destroyed. The incubation time of the eggs averaged 54 days, and the average
clutch size was 116 eggs.

The second purpose of the sea turtle protection program is to increase
the percentage of hatchlings that survive. Volunteers are trained and certified. When they find a nest that is too close to the water line or not buried
deep enough, they are authorized to relocate it to safer ground. Very carefully, a handful of sand at a time, they dig into the nest cavity. Then,
keeping each egg up-side up, they put it in a bucket, dig the new nest in
a better location, and place the eggs in the new nest in the exact order in
which they were extracted. On beaches where nests are prone to predation, volunteers will stake a protective fence over nests. Once the eggs
have hatched, the volunteers collect additional data by digging open every
fourth nest. They record the number of hatched and unhatched eggs and
the number that hatched but died in the nest. In the dog days of August,
this is hot, dirty, smelly, yolky work. One day, Judy dug open a nest and
found twenty dead hatchlings. It was a heartbreaker.

But every now and then, at the bottom of a nest, she will come across
a baby turtle that is still alive but who, for some reason, could not make it
to the surface on its own.

And she saves it.

Judy has been a volunteer in South Carolina's sea turtle nest protection program for five years. I ask her why she does it.

"Well, at first it was the thrill of riding a four-wheeler down the beach at dawn. It was exciting coming on those big turtle tracks—the crawls. But the first time you dig up a hatched-out nest to count the eggs and you see one of those little babies alive that did not make it to the surface, it just melts your heart."

"How many turtles do you think you have saved over the last five years?"

"Oh, lots, maybe hundreds." Wistfully, she adds, "But I know that, realistically, only one in a thousand will survive."

Then she laughs. "But, in my mind, every baby turtle that I ever helped makes it. I don't know what happens to the rest of them. But I know mine make it."

Then we make plans for the dawn patrol the next morning. Judy tells me to bring water and bug spray.

Six AM. Blue-and-gold Maxwell Parrish sky. The team consists of Judy, myself, a woman named Sherri Johnson, and April Fludd, a seventh-grader who has been working on the sea turtle project for two summers now. Our section of beach is on Botany Bay Plantation—6,500 acres of beach, salt marsh, and old fields and live oak forests. It is one of the most pristine beaches on the Atlantic Coast, a jewel.

We pick up a couple of ATVs loaded with gear: several five-foot-long T-shaped probes used to locate the nests, marker flags, a couple of rolled-up mesh cages to keep the raccoons out, a GPS, a big bucket in case we have to relocate a nest, bright orange signs to warn beachgoers against messing with a turtle nest. Judy tells me to hop on the back of the green four-wheeler with her, and April gets on the red one with Sherri. Zoom—we are off. The sun is coming up over the marsh. The air has a little chill, and we only hear birds. Down the half-mile trail toward the ocean, across the salt marsh, a snowy egret is wading, an osprey cruising overhead, a family of wood storks foraging in the shallows.

Judy looks back and yells to me over the rumble of the ATV: "See why I come out here? This is my church." She is right. Dawn at this beach is as transcendent as sitting in the nave at Chartres staring at stained glass.

After making our way through a very buggy bog, we hit the sand and start cruising, looking for crawls. Other than a shrimp boat trawling 500 yards offshore and a flock of low-flying pelicans in tight formation, the beach is empty. The ocean is placid, like a farm pond, and the coast of Africa looks to be just over the horizon.

Bad luck. We race the four-wheelers all the way to where the North Edisto River separates Botany Island from Seabrook Island, a tony, gated golf-and-tennis community for wealthy retirees, but we don't see a crawl. Zip. Nada. We turn around empty handed. I am discouraged.

Then, good luck. On the way back, we run into Chris Salmonsen, a state wildlife biologist who is responsible for the section of beach next to Judy's. We start to talk and when I tell him I am interested in turtle volunteers, people like Judy and Sherri and April, he smiles and says he could not do his job without them. We agree to meet later and he will fill me in on the human-sea turtle relationship.

Chris, forty-six, has a background in environmental education and has worked with turtle volunteers in Texas, Florida, and South Carolina.

"Tell me about the turtle people."

Some of them, he says, usually the men, like racing ATVs up and down the beach. They usually don't last very long. Most of the serious volunteers are women.

He says, "A lot of the volunteers have a space in their lives that they need to fill and the turtles help them fill that space. It is like going to church on Sunday. It is their religion."

I mention that Judy had told me exactly the same thing.

Judy is different from some of the volunteers, Chris says. She has lots going on in her life. Saving turtles is just one of the things she does. She also runs a business and is an artist.

Chris goes on. "That's not true of all the volunteers. Some of them are completely obsessed. They wear turtle shirts all the time, and their house is covered with turtle stuff. They want everyone to know they are turtle people. It *is* their identity."

Then he tells me about the time he was having dinner with a woman who was a turtle rescuer in Texas. When he ordered the shrimp special, she broke out into tears. It turns out that the woman was in a fight with local shrimpers over the use of TEDs—turtle excluder devices—that help loggerheads to escape from shrimp nets. Despite the TEDs, she believed that every shrimp cocktail translated to a dead loggerhead.

The next day I am up at dawn again, this time patrolling with Chris and his volunteers: a college student named Rosa who goes out five days a week and a photographer named Marie who is carrying a telephoto lens as long as my arm. I hop on Rosa's ATV and we take off through the bog next to the marsh. Today, the bugs in the marsh trail are really bad—*African Queen* bad. They bite through our clothes; we swat constantly at them with our baseball caps. Rosa has a trickle of blood running down her leg.

We drive down the beach and hit the first crawl after only a quarter mile. Rosa spots it first, which means that she has to find the hole leading to the nest cavity. She takes one of the probes and goes to work. You use the probe to locate the narrow tunnel leading to the nest cavity. The sand in the tunnel is more loosely packed than the surrounding ground. Probing is a delicate process, requiring enough pressure to sink the tip of the probe into the sand. But if you put too much weight on it and hit the loose sand of nest hole, the probe will dive into the nest cavity and skewer some of the eggs. Probing takes a deft touch; roughly 10% of all egg losses are caused by the probes.

It takes fifteen minutes of careful repeated probing an area two feet in diameter before Rosa feels the loose sand of the tunnel. She hands me the probe so I can get a sense of what she was going for. First, I push the probe into the sand next to the tunnel. It is not going anywhere. I move it over an inch, push it down again and, bingo, the loose sand in the tunnel immediately gives. Chris gets on hands and knees and starts to dig with one hand. By the time he reaches the eggs, his entire arm is engulfed in sand. He tells me to stick my arm in the hole, to feel the egg. It is leathery like an alligator's egg and somehow seems alive.

We fill the tunnel back up with sand, and Chris flags the nest, notes the GPS coordinates, and logs the data into his notebook. He will enter it into the computer when he gets back to his office, and it will show up online within twelve hours. We jump back on our ATVs and Chris quickly spots another crawl. He grabs the probe. We go through the drill, and are off again. I am feeling a little giddy. We have found two nests and I've got the sea turtle fever.

Several weeks later, I caught up with Meg Hoyle, who coordinates the volunteer program on Botany Island. I wanted her take on what makes people like Judy, Rosa, Sherri, and April get up at dawn and dig nests, fight off biting flies when it is 100 degrees in the shade, and come home smelling like rotten eggs, just to help animals that you hardly ever see, creatures that will almost certainly be munched up by predators long before they are old enough to breed.

She said, "Sometimes I don't understand it myself. Some of the volunteers walk the beach morning after morning, and in an entire summer, they might come across twelve nests. On most mornings, they will never find a crawl and many of them will never see a single turtle all season. Yet they stick with it. They are looking for a connection with the natural world that we have gotten away from. We all need that connection with animals and the outdoors."

THE ANTHROZOOLOGY OF EVERYDAY LIFE

Meg is right. Most people feel the need to connect with animals and nature. But we have this need to varying degrees. There are not many Michael Mountains, people who will not kill the ants invading their kitchens. There are, however, scads of Judy Muzzis, people who have jobs and families, people who are doing what they can in small ways to connect with animals, and are not particularly bothered by inconsistencies in their interactions with other species. They don't agonize over whether one should throw a switch that would send a hypothetical train careening into an old man or a group of endangered chimpanzees. They don't care whether the correct route to animal liberation runs through Bentham or Kant. Nor do they feel guilty over the fact that they refuse to eat beef but wear leather shoes.

I have—mostly—come to accept my own hypocrisies. The yahoo within me says that it is better to let Tilly outside than keep her imprisoned in the house all day even though I know she occasionally kills a towhee or a chipmunk. The yahoo tells me that the exquisite taste of slow-cooked pit barbecue somehow justifies the death of the hog whose loin I am going to slather with a pepper-based dry rub.

Moral intuitions change, however, and sometimes the yahoo and I make new arrangements. I quit fishing when I no longer found satisfaction in yanking a brown trout from a mountain stream. I don't eat veal anymore, we buy local eggs, and I am willing to pay more for a "free-range" chicken because I prefer to think it led a better life than a standard Cobb 500. And when an old rooster fighter recently asked me if I would like to go with him to a five-cock derby in Kentucky, I said no thanks.

When I first started studying human-animal interactions I was troubled by the flagrant moral incoherance I have described in these pages—vegetarians who sheepishly admitted to me they ate meat; cockfighters who proclaimed their love for their roosters; purebred dog enthusiasts whose desire to improve their breed has created generations of genetically defective animals; hoarders who caused untold suffering to the creatures living in filth they claim to have rescued. I have come to believe that these sorts of contradictions are not anomalies or hypocrisies. Rather, they are inevitable. And they show we are human.

Kwame Anthony Appiah, director of the Center for Human Values at Princeton, is sometimes asked what he does for a living. When he replies that he is a philosopher, the next question is usually, "So, what's your philosophy?" His standard response is, "My philosophy is that everything is more complicated than you thought."

What the new science of anthrozoology reveals is that our attitudes, behaviors, and relationships with the animals in our lives—the ones we love, the ones we hate, and the ones we eat—are, likewise, more complicated than we thought.

ACKNOWLEDGMENTS

My deep gratitude to the many animal people, from activists to zoologists, who have let me peer into their worlds and answered my naïve questions about their connections with other species.

Among the individuals who provided feedback on sections of the book are Jonathan Balcombe, Mihaly Bartalos, Marc Bekoff, Alex Bentley, Candace Boan-Lenzo, Leo Bobadilla, Larry Carbone, Merritt Clifton, Jane Coburn, Chris Coburn, Karen Davis, Judy DeLoache, Bev Dugan, Leah Gomez, John Goodwin, Sam Gosling, Che Green, Joshua Greene, Katherine Grier, Liz Hodge, Leslie Irvine, Wes Jamison, Rebecca Johnson, Sarah Knight, Kathy Kruger, Laura Maloney, Lori Marino, Ádám Miklósi, Michael Mountain, Jim Murray, David Nieman, Emily Patterson-Kane, Scott Plous, Andrew Rowan, Kathy Rudy, Boria Sax, Ken Shapiro, Michael Schafer, Harriet Shields, Jeff Spooner, Craig Stanford, Samantha Strazanac, Mickey Randolph, Lee Warren, Erin Williams, Richard Wrangham, Steven Wise, Clive Wynne, and Steve Zawistowski. Morgan Childers, Sue Grider, and the staff of the Hunter Library at Western Carolina University came to the rescue in terms of references and manuscript preparation.

For many years, my scholarly home has been the International Society for Anthrozoology, an enthusiastic group of researchers who transcend traditional academic pigeon holes. I am especially indebted to Arnie Arluke, Alan Beck, Ben and Lynette Hart, Anthony Podberscek, and James Serpell for their encouragement and their contributions to the study of human-animal relationships. I owe much to the ethologist Gordon Burghardt, my longtime mentor and friend. Paul Rozin and Jon Haidt, psychologists for whom thinking outside the box seems to come easy, have deeply influenced my views on the psychology of morality.

David Henderson and Chris Diehm helped clarify the intricacies of animal rights philosophies for me. Rob Bass and Gail Dean read most chapters, and this book is much better for their insight and critiques. Joyce Moore of City Lights Bookstore in Sylva, North Carolina convinced me that publishers might actually be interested in a book on why it is so hard to think straight about animals. When in need of literary inspiration, I turned to Harry Greene, Robert Sapolsky, Elmore Leonard, and Merle Haggard.

This book would not have been published except for the efforts of my agents Jennifer Gates and Rachel Sussman. Jennifer saw potential in my fledgling idea (about which another agent told me "No one would want to read about that"). Rachel worked tirelessly to shape the proposal that formed the core of the book, always gently drumming into my head that a good book is more than a collection of interesting anecdotes. The editorial staff at Harper has been a dream team. Executive editor Gail Winston was the perfect task master. She reined me in when I needed it and became the little voice in my head that said, "Always remember that your readers are smart." The sentences are better for the skilled hand of Amy Vreeland who also asked the right questions. Jason Sack was adept at shepherding the manuscript through the production process. Publisher Jonathan Burnham immediately understood the book's message and had the insight to know that it needed an additional chapter.

I could not ask for better colleagues than the faculty of the Psychology Department at Western Carolina University. They continue to be extraordinarily tolerant when I barge into their offices waving the latest dog breed graph or railing about a new research paper I have run across. For twenty years, Bruce Henderson and David McCord have read drafts of my papers and let me know when I made a wrong turn. Much of my research has been conducted in collaboration with graduate and undergraduate students at Western Carolina University. I hope they had as much fun as I did.

Writing a book can make you crazy, and I got by with a little help from my friends—actually lot of help. My longtime kayaking pals helped me keep things in perspective by reminding me when I needed to go with the flow. For nearly fifteen years, I have recharged my batteries every Tuesday night playing old time mountain music at Guadalupe's Restaurant and Spring Street Café with a cast of great musicians led by fiddler extraordinaire Ian Moore. And thanks to Jen and Faye for the (cruelty free) goat tacos and beer. Special thanks go to Mac Davis who I met when we moved into a small farmhouse up Sugar Creek. He was across the road plowing his tobacco field with a mule. Thirty-five years later, I am still running ideas by him, including many contained in this book.

My family has been extraordinarily supportive during the months I spent buried in a dingy basement office amid piles of reprints and stale cups of cold coffee. My brother, sister, and mother were a source of constant encouragement. My collaborator-for-life is my wife, Mary Jean, who early in our relationship proved her mettle by helping demonstrate that angry mother alligators would actually attack a human intruder in defense of their young. More recently, she uncomplainingly read every sentence in this book a least a half dozen times and has generally kept me sane. She is the best. Our children, Adam, Katie, and Betsy, all fine writers, cheerfully critiqued chapters. I could always rely on them for honest advice like, "Dad, this sentence sucks."

Finally, a Crunchy Salmon Treat to Tilly, who spent many a drowsy afternoon lying in a rocking chair, keeping me company and watching me write, occasionally meowing so I would rub her belly, reminding me why we bring animals into our lives.

RECOMMENDED READING

ANTHROZOOLOGY

Arluke, A., and Bogdan, R. (2010). *Beauty and the Beast: Human-Animal Relationships as Revealed in Real Post Cards.* Syracuse, NY: Syracuse University Press.

Bekoff, M., ed. (2007). *Encyclopedia of Human-Animal Relationships.* Westport, CT: Greenwood Press.

Kalof, L. (2007). *Looking at Animals in Human History.* London: Reaktion Books.

Ritvo, H. (1989). *The Animal Estate: The English and Other Creatures in the Victorian Age.* Cambridge, MA: Harvard University Press.

Serpell, J. (1996). *In The Company of Animals: A Study of Human-Animal Relationships.* Cambridge, UK: Cambridge University Press.

PSYCHOLOGY AND HUMAN-ANIMAL INTERACTIONS

Bulliet, R. W. (2005). *Hunters, Herders, and Hamburgers: The Past and Future of Human-Animal Relationships.* New York: Columbia University Press.

Haraway, D. J. (2008). *When Species Meet.* Minneapolis: University of Minnesota Press.

Melson, L. G. (2001). *Why the Wild Things Are: Animals in the Lives of Children.* Cambridge, MA: Harvard University Press.

Wilson, E. O. (1992). *Biophilia: The Human Bond with Other Species.* Cambridge, MA: Harvard University Press.

PETS

Anderson, P. E. (2008). *The Powerful Bond between Pets and People.* Westport, CT: Praeger.

Grier, K. C. (2006). *Pets in America: A History*. Chapel Hill: University of North Carolina Press.

Irvine, L. (2004). *If You Tame Me: Understanding Our Connection with Animals*. Philadelphia, PA: Temple University Press.

Zawistowski, S. (2008). *Companion Animals in Society*. Clifton Park, NY: Thompson Delmar Learning.

DOGS

Coppinger, R., and Coppinger, L. (2002). *Dogs: A New Understanding of Canine Origin, Behavior, and Evolution*. Chicago: University of Chicago Press.

Horowitz, A. (2009). *Inside of a Dog: What Dogs See, Smell, and Know*. New York: Scribner.

McConnell, P. B. (2002). *The Other End of the Leash: Why We Do What We Do around Dogs*. New York: Ballantine Books.

Miklósi, A. (2007) *Dog Behavior, Evolution, and Cognition*. New York: Oxford University Press.

Sanders, C. (1999). *Understanding Dogs: Living and Working with Canine Companions*. Philadelphia, PA: Temple University Press.

Schaffer, M. (2006). *One Nation under Dog*. New York: Henry Holt.

Serpell, J. (1995). *The Domestic Dog: Its Evolution, Behaviour, and Interactions with People*. Cambridge, UK: Cambridge University Press.

GENDER

Baron-Cohen, S. (2003). *The Essential Difference: The Truth about the Male and Female Brain*. New York: Basic Books.

Donovan, J., and Adams, C. J. (1996). *Beyond Animal Rights: A Feminist Caring Ethic for the Treatment of Animals*. New York: Continuum.

Luke, B. (2007). *Brutal: Manhood and the Exploitation of Animals*. Chicago: University of Illinois Press.

MEAT

Adams, C. J. (2000). *The Sexual Politics of Meat: A Feminist-Vegetarian Critical Theory*. New York: Continuum Publishing Group.

Eisnitz, G. A. (2007). *Slaughterhouse: The Shocking Story of Greed, Neglect, and Inhumane Treatment Inside the U.S. Meat Industry*. Amherst, NY: Prometheus Books.

Foer, J. S. (2009). *Eating Animals*. New York: Little, Brown and Company.

Maurer, D. (2002). *Vegetarianism: Movement or Moment?* Philadelphia, PA: Temple University Press.

Pollan, M. (2006). *The Omnivore's Dilemma: A Natural History of Four Meals*. New York: Penguin.

Singer, P., and Mason, J. (2006). *The Ethics of What We Eat: Why Our Food Choices Matter*. Emmaus, PA: Rodale Press.

CHICKENS

Davis, K. (2009). *Prisoned Chickens, Poisoned Eggs: An Inside Look at the Modern Poultry Industry*. Summertown, TN: Book Publishing Company.

Dundes, A. (1994). *The Cockfight: A Casebook*. Madison: University of Wisconsin Press.

Smith, P., and Daniel, C. (2000). *The Chicken Book*. Athens: University of Georgia Press.

MICE AND SCIENCE

Birke, L., Arluke, A., and Michael, M. (2008) *The Sacrifice: How Scientific Experiments Transform Animals and People*. West Lafayette, IN: Purdue University Press.

Blum, D. (1995). *The Monkey Wars*. New York: Oxford University Press.

Morrison, A. R. (2009). *An Odyssey with Animals: A Veterinarian's Reflections on the Animal Rights and Welfare Debate*. Oxford, UK: Oxford University Press.

Rudacille, D. (2000). *The Scalpel and the Butterfly: The War between Animal Research and Animal Protection*. New York: Farrar, Straus, and Giroux.

CONSISTENCY AND ETHICS

Balcombe, J. (2010). *Second Nature: The Inner Lives of Animals*. New York: Palgrave MacMillan.

Bekoff, M., and Pierce, J. (2009). *Wild Justice: The Moral Lives of Animals*. Chicago: University of Chicago Press.

Hauser, M. D. (2006). *Moral Minds: How Nature Designed Our Universal Sense of Right and Wrong*. New York: HarperCollins.

Kazez, J. K. (2010). *Animalkind: What We Owe to Animals*. Malden, MA: Wiley-Blackwell.

Singer, P. (1990). *Animal Liberation: A New Ethics for Our Treatment of Animals*. New York: Avon.

Wynne, C. D. L. (2004). *Do Animals Think?* Princeton, NJ: Princeton University Press.

NOTES

1 *Marc Bekoff* speech to the Farm Sanctuary Hoe Down (www.farmsanctuary. org) in Orland, California, on May 16, 2009.

6 *all members of the family Felidea eat flesh for a living* Cats, unlike dogs, need meat to stay healthy. For a comparison between the nutritional needs of dogs and cats, see Legrand-Defretin, V. (1994). Differences between cats and dogs: A nutritional view. *Proceedings of the Nutrition Society,* 53, 15–24.

6 *cats are recreational killers* Crook, K. R., & Soule, M. E. (1999). Mesopredator release and avifaunal extinction in a fragmented system. *Nature,* 400, 563–566. Also see Woods, M., McDonald, R. A., & Harris, S. (2003). Predation of wildlife by domestic cats (*Felis catus*) in Great Britain. *Mammal Review,* 33(2), 174–188.

6 *A group of Kansas cat owners* Stuchhury, B. (2007) *Silence of the songbirds: How we are losing the world's songbirds and what we can do to save them.* Toronto: HarperCollins.

8 *bone in their penis* The penis bone is called the *baculum* or *os penis.* Penis bones are found in many species of mammals, including primates. In different animals, these bones vary widely in size and shape. Why some species have big ones, others small ones, and some, including humans, none at all remains an evolutionary mystery. Ramm, S. A. (2007). Sexual selection and genital evolution in mammals: A phylogenetic analysis of baculum length. *American Naturalist,* 169, 360–369.

9 *Four and a half million Americans are bitten by dogs* Sacks, J. J., Kresnow, M., & Houston, B. (1996). Dog bites: How big a problem? *Injury Prevention,* 2(1), 52–54. Sacks, J. J., Sinclair, L., Gilchrist, J., Golab, G. C., & Lockwood, R. (2000). Breeds of dogs involved in fatal human attacks in the United States between 1979 and 1998. *Journal of the American Veterinary Medical Association,* 217(6), 836–840.

9 *canine train wreck* Serpell, J. A. (2003). Anthropomorphism and anthropo-

morphic selection: Beyond the "cute response." *Society & Animals,* 11(1), 83–100.

11   *back issues of sleazy supermarket tabloids* Herzog, H. A., & Galvin, S. L. (1992). Animals, archetypes, and popular culture: Tales from the tabloid press. *Anthrozoös,* 5, 77–92.

11   *the "troubled middle"* Donnelley, S. (1989). Speculative philosophy, the troubled middle, and the ethics of animal experimentation. *Hastings Center Report,* 19, 15–21.

15   *Clifton Flynn, sociologist* Flynn, C. (2008). *Social creatures: A human and animal studies reader.* Brooklyn, NY: Lantern Books. (p. xiv).

17   *several journals that publish our research* The most prominent journals devoted to human-animal relationships are *Anthrozoös* and *Society and Animals.*

17   *the International Society for Anthrozoology* The organization's Web site is www.isaz.net.

18   *Animal-assisted therapy* For an overview of recent research in animal-assisted therapy, see Fine, A. H. (ed.) (2008). *Handbook on animal assisted-therapy: Theoretical foundation and guidelines for practice.* New York: Elsevier.

18   *forty-nine published studies of the effectiveness of AAT* Nimer, J., & Lundahl, B. (2007). Animal-assisted therapy: a meta-analysis. *Anthrozoös,* 20 (3), 225–238.

19   *the claims made about their curative powers are over the top.* See, for example, Cochrane, A., & Callen, K. (1992). *Dolphins and their power to heal.* Rochester, VT: Healing Arts Press.

20   *Hawthorne Works* According to Jim Goodwin, an expert on the history of psychology, the Hawthorne studies had a little-known political subplot. The real purpose of the study may have been for management to portray factory workers as happy with their jobs and thus impede unionization. Goodwin, C. J. (2008). *A history of modern psychology* (3rd ed.). New York: J. Wiley.

20   *A group of German researchers* Brensing, K., Linke, K., & Todt, D. (2003). Can dolphins heal by ultrasound? *Journal of Theoretical Biology,* 225(1), 99–105.

21   *They found that every one of them was methodologically flawed* Marino, L., & Lilienfeld, S. O. (1998). Dolphin-assisted therapy: Flawed data, flawed conclusions. *Anthrozoös,* 11(4), 194–200. Marino, L., & Lilienfeld, S. O. (2007). Dolphin-assisted therapy: More flawed data and more flawed conclusions. *Anthrozoös,* 20, 239–249. Tracy Humphries reached a similar conclusion. Humphries, T. L. (2003). Effectiveness of dolphin-assisted therapy as a behavioral intervention for young children with disabilities. *Bridges,* 1(6), 1–9.

22   *suffered traumatic injuries* Mazet, J. A., Hunt, T. D., & Zoccardi, M. H. (2004). *Assessment of the risk of zoonotic disease transmission to marine mammal workers and the public.* Final report. United States Marine Mammal Commission, RA No. K005486–01.

23   *In a letter released by the Aruba Marine Mammal Foundation* Smith, B. A.

(2007). *Letter of DAT founder.* (Retrieved August 28, 2008.) www.arubammf. com/truth_about_dolphin_assisted_therapy

23   *Save your money; save a dolphin* For an engaging overview of the relationships between humans and dolphins, see Kudzinski, K. M., & Frohoff, T. (2008). *Dolphin mysteries: Unlocking the secrets of communication.* New Haven, CT: Yale University Press.

24   *do people really look like their dogs* Coren, S. (1999). Do people look like their dogs? *Anthrozoös,* 12, 111–114. Roy, M. M., & Christenfeld, N. J. (2004). Do dogs resemble their owners? *Psychological Science,*15(5), 361–363. Payne, C., & Jaffe, K. (2005). Self seeks like: Many humans choose their dog pets following rules used for assortative mating. *Journal of Ethology,* 23(1), 15–18; Devlin, K. (April 3, 2009) Dog owners do look like their pets, say psychologists. *Telegraph* (London); Nakajima, S., Yamamoto, M., & Yoshimoto, N. (2009). Dogs look like their owners: Ratings of dogs with racially homogenous owner portraits. *Anthrozoös,* 22(2), 173–181.

27   *Some of our tastes reveal aspects of our personality, but others do not* Gosling's approach to research is a creative mix of evolutionary psychology, personality theory, and animal behavior. You can keep up with his latest research at the Gozlab Web site: homepage.psy.utexas.edu/homepage/faculty/gosling/index. htm.

27   *Gosling and Anthony Podberscek* Podberscek, A. L., & Gosling, S. D. (2005). Personality research on pets and their owners: Conceptual issues and review. In A. Podberscek, E. S. Paul, & J. A. Serpell (eds.), *Companion animals and us: Exploring the relationships between people and pets* (pp. 143–167). Cambridge, UK: Cambridge University Press.

28   *527 cat people had taken the personality test* Gosling, S. Sandy, C. J., & Potter, J. (in press). Personalities of self-identified "dog people" and "cat people." *Anthrozoös*

29   *Some scientists believe that roots of cruelty lie in our evolutionary history* Victor Nell, for example, argues that cruelty is the natural extension of human predatory instincts. Nell, V. (2006). Cruelty's rewards: The gratifications of perpetrators and spectators. *Behavioral and Brain Sciences,* 29, 211–224.

29   *The anthropologist Margaret Mead wrote* This quote is cited on page 80 of Lockwood, R., & Hodge, G. R. (1998). The tangled web of animal abuse: The links between cruelty to animals and human violence. In R. Lockwood & F. R. Ascione (eds.), *Cruelty to animals and interpersonal violence* (pp. 77–82). West Lafayette, IN: Purdue University Press.

30   *Others, however, are not so sure* For an introduction to this topic see the articles in Lockwood, R., & Ascione, F. R. (1998). *Cruelty to animals and interpersonal violence: Readings in research and application.* West Lafayette, IN: Purdue University Press. For a review of current research in this area see Ascione, F. R., & Shapiro, K. (2009). People and animals, kindness and cruelty: Research directions and policy implications. *Journal of Social Issues,* 65, 569–587.

30   *Alan Felthous, a psychiatrist and Stephen Kellert* Felthous, A. R., & Kellert, S. R. (1986). Violence against animals and people: Is aggression against living creatures generalized? *The Bulletin of the American Academy of Psychiatry and the Law,* 14(1), 55–69.

30   *"I beat a puppy, I believe simply from enjoying the power."* This quote is found in Blakemore, C. (February 12, 2009) Darwin understood the need for animal tests. *Times (London)* Online. www.timesonline.co.uk/tol/comment/columnists/guest_contributors/article5711912.ece

31   *presentations by Link advocates often begin with tales of tragedy* See, for example, Lockwood, R., & Hodge, G. R. (1998). The tangled web of animal abuse: The links between cruelty to animals and human violence. In R. Lockwood & F. R. Ascione (ed.), *Cruelty to animals and interpersonal violence* (pp. 77–82). West Lafayette, IN: Purdue University Press.

31   *354 cases of serial murders* The researchers in this study found that 21% of the 354 serial killers had a history of animal abuse. Most studies of male college students have also found animal abuse in the 20% to 30% range. The serial killer data is reported in Wright, J., & Hensley, C. (2003). From animal cruelty to serial murder: Applying the graduation hypothesis. *International Journal of Offender Therapy and Comparative Criminology,* 47, 71–88.

31   *school shootings* The claim that all school shooters have a history of animal abuse is made by the National District Attorneys Association (communities.justicetalking.org/blogs/day17/archive/2007/08/14/animal-abuse-its-association-with-other-violent-crimes.aspx). For the actual profile of school shooters, see Vossekuil, B., Fein, R., Reddy, M., Borum, R., & Modzeleski, W. (2000). *The final report and findings of the Safe School Initiative: Implications for the prevention of school attacks in the United States.* Washington, DC: United States Secret Service.

32   *A stronger version of Link thinking is the violence graduation hypothesis* This idea is sometime referred to as the progression thesis. Beirne, P. (2004). From animal abuse to interhuman violence? A critical review of the progression thesis. *Society and Animals,* 12(1), 39–65.

32   *A group of researchers led by Arnold Arluke* Arluke, A., Levin, J., Luke, C., & Ascione, F. (1999). The relationship of animal abuse to violence and other forms of antisocial behavior. *Journal of Interpersonal Violence,* 14(9), 963–975.

33   *the numbers do not support the idea* Patterson-Kane, E. G., & Piper, H. (2009). Animal abuse as a sentinel for human violence: A critique. *Journal of Social Issues.* Lea, S. R. G. (2007). *Delinquency and animal cruelty: Myths and realities about social pathology.* New York: LFB Scholarly Publishing, LLC.

33   *The students he interviewed had poisoned fish* Arluke, A. (2002). Animal abuse as dirty play. *Symbolic Interaction,* 25(4), 405–430.

33   *One recent study* Gupta, M. E. (2006). *Understanding the links between intimate partner violence and animal abuse: Prevalence, nature, and function.* Unpublished doctoral dissertation, University of Georgia, Athens, GA.

34 *A growing number of researchers* See, for example, Piper, H. (2003). The linkage of animal abuse with interpersonal violence: A sheep in wolf's clothing? *Journal of Social Work,* 3(2), 161–176. Irvine, L. (2008). Delinquency and animal cruelty: Myths and realities about social pathology. *Contemporary Sociology: A Journal of Reviews,* 37(3), 267–268. Taylor, N., & Signal, T. (2008). Throwing the baby out with the bathwater: Towards a sociology of the human-animal abuse 'Link'? *Sociological Research Online,* 13(1). www.socresonline. org.uk/13/1/2.html.

35 *"Animals are good to think with."* Levi-Strauss, C. (1966). *The savage mind.* Chicago: University of Chicago Press.

37 *It's easier to empathize* Greene, E S. (1995). Ethnocatetgories, social intercourse, fear, and redemption: Comment on Laurent. *Society and Animals,* 3(1), 79–88.

37 *cuteness doesn't count* Cohen, R. (July 19, 2009). The ethicist: Nesting blues. *New York Times.*

38 *decisions of ant colonies* Edwards, S. C., & Pratt, S. C. (2009). Rationality in collective decision making by ant colonies. *Proceedings of the Royal Society: B.* 276(1673) 3655–3671.

39 *biophilia* Wilson, E. O. (1984). *Biophilia.* Cambridge, MA: Harvard University Press. The study of infants is found in Deloache, J. S., & Pickard. (in press). How very young children think about animals. The animate monitoring hypothesis is found in New, H., Cosmides, L., & Tooby, J. (2007). Category-specific attention for animals reflects ancestral priorities, not experience. *PNAS,* 104(42), 16598–16603.

39 *Bambi is the classic example of how easily we are manipulated by instinctive baby releasers* Lutts, R. H. (1992). The trouble with Bambi: Walt Disney's *Bambi* and the American vision of nature. *Forest and Conservation History,* 36, 160–171. Cartmill, M. (1993). *A view to death in the morning.* Cambridge, MA: Harvard University Press.

40 *Stephen Jay Gould, the late Harvard biologist who traced Mickey's evolution, said it best* Gould, S. J. (1979). Mickey Mouse meets Konrad Lorenz. *Natural History,* 88(5), 30–36.

43 *a rattlesnake bite on the end of his tongue* Gerkin, R., Sergent, K. C., Curry, S. C., Vance, M., Nielsen, D. R., & Kazan, A. (1987). Life-threatening airway obstruction from rattlesnake bite to the tongue. *Annals of Emergency Medicine,* 16(7), 813–816.

43 *snake fears* for research on snake fears, see Öhman, A., & Mineka, S. (2003). The malicious serpent: Snakes as a prototypical stimulus for an evolved module of fear. *Current Directions in Psychological Science,* 12(1), 5–9. LoBue, V., & DeLoache, J. S. (2008). Detecting the snake in the grass: Attention to fear-relevant stimuli by adults and young children. *Psychological Science,* 19(3), 284–289. DeLoache, J. S., & LoBue, V. (2009). The narrow fellow in the grass: Human infants associate snakes and fear. *Developmental Science,* 12(1),

201–207. Diamond, J. (1993). New Guineans and their natural world. In Kellert & Wilson (eds.), *The biophilia hypothesis* (pp. 251–271). Washington, DC: Island Press. Burghardt, G. N., Murphy, J. B., Chiszar, D., & Hutchins, M. (2009). Combating ophidophobia: Origins, treatment, education and conservation tools. In S. J. Mullin & R. A. Seigel (eds). *Snakes: Ecology and conservations*. Ithaca, NY: Comstock Publishing.

44 *"Biophilia," he wrote, "is not a single instinct"* Wilson, E. O. (1993). Biophilia and the conservation ethic. In S. R. Kellert & E. O. Wilson (eds.) *The biophilia hypothesis* (p. 31–41). Washington, DC: Island Press. (p. 31).

44 *Animal words permeate human language* The classic essay on the influence of animal words on language is Leach, E. R. (1964). Anthropological aspects of language: Animal categories and verbal abuse. In E. Lenneberg (ed.), *New directions in the study of language* (pp. 23–63). Cambridge, MA: MIT Press.

47 *fell in love with a beagle puppy* This case is described in Herzog, H. (2002). Ethical aspects of the relationship between humans and research animals. *ILAR Journal*, 43, 27–32. All of the laboratory animal technicians that I interviewed for this research had adopted research animals.

47 *The human propensity for categorizing animals* Grief, M. L., Nelson, D. G. K., Keil, F. C., & Gutierrez, T. (2006). What do children want to know about animals and artifacts? *Psychological Science*, 17, 455–459.

47 *blind from birth* Mahone, B. Z., Anzellotti, S., Schwarzbach, J., Zampini, M., & Caramazza, A. (2009). Category-specific organization of the human brain does not require visual experience. *Neuron*, 63, 397–405.

47 *human brain evolved to specialize* The idea of domain-specific knowledge is controversial among neuroscientists. For contrasting views see Caramazza, A., & Shelton, J. R. (1998). Domain-specific knowledge system in the brain: The animate–inanimate distinction. *Journal of Cognitive Neuroscience*, 10, 1–34 and Gerlach, C. (2007). A review of functional imaging studies on category specificity. *Journal of Cognitive Neuroscience*, 19, 296–314.

48 *"sociozoologic scale"* Arluke, A., & Sanders, C. (1996). *Regarding animals*. Philadelphia, PA: Temple University Press.

48 *In Japan, attitudes toward creepy-crawlies are more complex* Laurent, E. L. (2001). Mushi: For youngsters in Japan, the study of insects has been both a fad and a tradition. *Natural History*, 110(2), 70–75.

49 *anthrozoologist James Serpell* Serpell, J. A. (2004) Factors influencing human attitudes to animals and their welfare. *Animal Welfare*, 13, S145–151.

49 *How Pigeons Became Rats* Jerolmack, C. (2008). How pigeons became rats: The cultural-spatial logic of problem animals. *Social Problems*, 55, 72–94.

51 *the croc's death as "emotionally satisfying yet thoroughly irrational."* No author. (September 12, 1977). Kill the crocodile. *New York Times* (p. 32).

51 *The debate over whether human morality is based on emotion or reason* For an excellent overview of moral psychology, see Hauser, M. D. (2006). *Moral*

minds: How nature designed our universal sense of right and wrong. New York: HarperCollins.

52 *Shelley Galvin and I used this method to investigate how people make decisions about the use of animals in research* Galvin, S. L., & Herzog, H. A. (1992). The ethical judgment of animal research. *Ethics & Behavior,* 2(4), 263–286.

52 *Haidt's theory* Haidt, J. (2001). The emotional dog and its rational tail: A social intuitionist approach to moral judgment. *Psychological Review,* 108(4), 814–834.

53 *Paul Rozin calls disgust the moral emotion* Rozin, P., Haidt, J., & McCauley, C. R. (1999). Disgust: The body and soul emotion. In T. Dalgleish & M. Power (eds.), *Handbook of cognition and emotion* (pp. 429–445). Chichester, UK: Wiley.

55 *causes the emotional processing centers of the brain to light up while the impersonal version (throwing the switch) does not* For examples of Greene's research on the neuroanatomy of morality see Greene, J., & Haidt, J. (2002). How (and where) does moral judgment work? *Trends in Cognitive Sciences,* 6(12), 517–523.

55 *University of California psychologist Lewis Petrinovich* Petrinovich, L., O'Neill, P., & Jorgenson, M. (1993). An empirical study of moral intuitions: Toward an evolutionary ethics. *Journal of Personality and Social Psychology,* 64(3), 467–478.

55 *Marc Hauser, director of the Cognitive Evolution Laboratory* Hauser, M. D. (2006). *Moral minds: How nature designed our universal sense of right and wrong.* New York: HarperCollins.

57 *Richard Bulliet* Bulliet, R. W. (2005). *Hunters, herders, and hamburgers: The past and future of human-animal relationships.* New York: Columbia University Press.

57 *Answer these two questions* The bat and ball question still drives me crazy. Think of it this way: If the bat cost $1 and the ball cost 10 cents, then the bat cost 90 cents more than the ball, not a dollar more. However, if the bat costs $1.05 and the ball costs 5 cents, then the bat costs a dollar more than the ball, and together they add up to $1.10. These examples are from Plous, S. (1993). *The psychology of judgment and decision making.* New York: McGraw-Hill.

57 *Our predilection for revenge* Sunstein, C. R. (October 2002). Hazardous heuristics. *University of Chicago Law & Economics Olin Law and Economics Working Paper No. 165* and *University of Chicago Public Law Research Paper No. 33.*

58 *heuristic is called framing* The scientific literature on heuristic is vast. For examples, see Hallinan, J. T. (2009). *Why we make mistakes.* New York: Broadway Books. Marcus, G. (2008). *Kluge: The haphazard construction of the human mind.* New York: Houghton Mifflin. Kahneman, D., & Frederick, S. (2002). Representativeness revisited: Attribute substitution in intuitive judgment. In T. Gilovick, D. Grifin, & D. Kahneman (eds.), *Heuristics and biases: The psychology of intuitive judgment* (pp. 49–81). Cambridge, UK: Cambridge University Press.

58   *the Nazi animal protection movement* For an excellent description of the Nazi animal protection movement see Arluke, A., & Sax, B. (1992). Understanding Nazi animal protection and the Holocaust. *Anthrozoös*, 5(1), 6–31. Sax, B. (2000). *Animals in the Third Reich: Pets, scapegoats, and the Holocaust*. New York: Continuum International Publishing Group.

59   *contemporary animal activists don't relish the idea that Adolf Hitler was a fellow traveler* See, for example, Berry, R. (2004). *Hitler: Neither vegetarian nor animal lover*. Brooklyn, NY: Pythagorean Books.

61   *how children and adults responded to an AIBO compared to a real dog* Melson, G. F., Kahn, P., Beck, A., Friedman, B., & Edwards, N. (2009). Robotic pets in human lives: Implications for the human-animal bond and for human relationships with personified technologies. *Journal of Social Issues*, 65, 545–567.

61   *AIBO could also alleviate human loneliness* Banks, M. R., Willoughby, L. M., & Banks, W. A. (2008). Animal-assisted therapy and loneliness in nursing homes: Use of robotic versus living dogs. *Journal of the American Medical Directors Association*, 9(3), 173–177.

62   *Serpell eloquently lays out the moral issues* Serpell, J. A. (1996). *In the company of animals: A study of human-animal relationships*. Cambridge, UK: Cambridge University Press. (p. 177).

63   *Researchers at the University of Portsmouth* Morris, P. H. (2008). Secondary emotions in non-primate species? Behavioural reports and subjective claims by animal owners. *Cognition & Emotion*, 22(1), 3–20.

63   *"What Is It Like to Be a Bat?"* Nagel, T. (1974). What is it like to be a bat? *Philosophical Review*, 4, 435–450.

64   *what Gordon Burghardt calls critical anthropomorphism* Burghardt, G. M. (1991). Cognitive ethology and critical anthropomorphism: A snake with two heads and hognose snakes that play dead. In C. A. Ristau (ed.), *Cognitive ethology: The minds of other animals* (pp. 53–90). Hillsdale, NJ: Lawrence Erlbaum Associates. Not all animal behaviorists are enthusiastic about the emergence of anthropomorphism as an ethological research tool. See, for example, Wynne, C. D. L. (2004). The perils of anthropomorphism. *Nature*, 428, 606.

67   *Assume that animal companions* Holbrook, M. B. (2008). Pets and people: Companions in commerce. *Journal of Business Research*, 61, 546–552.

67   *Antoine, a young Frenchman* Gueguen, N., & Ciccotti, S. (2008). Domestic dogs as facilitators in social interaction: An evaluation of helping and courtship behaviors. *Anthrozoös*, 21, 339–349.

72   *The historian Keith Thomas argues that pets are animals that are allowed in the house* Thomas, K. (1983). *Man and the natural world: A history of modern sensibility*. New York: Pantheon.

73   *University of Pennsylvania anthrozoologist James Serpell* Serpell, J. A. Petkeeping and animal domestication: A reappraisal. In Clutton-Brock, J. (ed.) (1989). *The walking larder: Patterns of domestication, pastorialism and predation* (pp. 10–21). London: Unwin Hyman.

73 *most animals in American homes* Grier, K. C. (2006). *Pets in America: A history.* Chapel Hill: University of North Carolina Press.

74 *more like slavery* Irvine, L. (2004). Pampered or enslaved? The moral dilemmas of pets. *International Journal of Sociology and Social Policy, 24,* 5–17. (p. 14).

75 *the amount of money we dole out* American Pet Products Association. (2009). *Industry trends and statistics.* Retrieved August 30, 2009, from americanpet products.org/press_industrytrends. asp. Brady, D., & Palmeri, C. (August 6, 2007). The pet economy. *Business Week.*

77 *nineteenth-century France.* Kete, K. (1995). *The beast in the boudoir: Petkeeping in nineteenth-century Paris.* Berkeley: University of California Press.

77 *a trend that Michael Silverstein and Neil Fiske call trading up* Silverstein, M. J., & Fiske, N. (2003). *Trading up: Why consumers want new luxury goods—and how companies create them.* New York: Portfolio.

77 *answer is yes* Some aspects of the pet industry have been more affected by the economic downturn than others. For example, according to James Serpell, the University of Pennsylvania Veterinary Hospital has seen a dramatic drop in expensive medical procedures for dogs and cats in the last two years.

78 *advice to corporations trying to tap into the lucrative pet products marketplace* Holbrook, M. B. (2008). Pets and people: Companions in commerce. *Journal of Business Research, 61,* 546–552.

78 *benefits they derived from their relationships* Herzog, H., Kowalski, R., Burgner, M., & Dunegon, C. (May 2003). Are pets really friends? Perceived benefits of relationships with companion animals. Paper presented at the meeting of the American Psychological Society, Atlanta, GA.

78 *Human friends were better than animal companions* Serpell, J. (1989). Humans, animals, and the limits of friendship. In Porter, R., & Tomaselli, S. (eds.), *The dialectics of friendship* (pp. 111–129). London: Routledge.

79 *not particularly attached to their animals* Johnson, T. P., Garrity, T. F., & Stallones, L. (1992). Psychometric evaluation of the Lexington Attachment to Pets Scale. *Anthrozoös, 5,* 160–175. A lot of research has focused on people who are attached to their pets. However, I don't know of any systematic investigations of the millions of people who live with animals whom they dislike.

79 *pet attachment drops a notch* Poresky, R. H., & Daniels, A. M. (1998). Demographics of pet presence and attachment. *Anthrozoös, 11,* 236–241.

81 *owning a pet made a big difference in their survival rates* Friedmann, E., Katcher, A. H., Lynch, J. J., & Thomas, S. A. (1980). Animal companions and one-year survival of patients after discharge from a coronary care unit. *Public Health Reports, 95*(4), 307–312. Erica later replicated these effects, this time with more subjects so that the effects of owning dogs and cats could be evaluated separately. While owning a dog had a big effect on survivorship, owning a cat had no effect. See Friedmann, E., & Thomas, S. (1995). Pet ownership, social support, and one-year survival after acute myocardial infarction in the

cardiac arrhythmia suppression trial (CAST). *American Journal of Cardiology*, 76, 1213–1217.

81   *the notion that pets have beneficial effects on human health and well-being*   For reviews of this research see Friedmann, E., Thomas, S., & Eddy, T. (2000). Companion animals and human health: Physical and cardiovascular influences. In Podberscek, A. L., Paul, E. S., Serpell, J. A. (ed.), *Companion animals and us: Exploring the relationships between people and pets* (pp. 125–142). Cambridge, UK: Cambridge University Press. Wells, D. L. (2009). The effects of animals on human health and well-being. *Journal of Social Issues*, 65, 523–543. Beck, A. M., & Katcher, A. H. (2003). Future directions in human-animal bond research. *American Behavioral Scientist*, 47(1), 79–93. Barker, S. B., & Wolden, A. R. (2008). The benefits of human-companion animal interaction: A review. *Journal of Veterinary Medical Education.* 35, 487–495. Headey, B., & Grabka, M. M. (2007). Pets and human health in Germany and Australia: National longitudinal results. *Social Indicators Research*, 80(2), 297–311.

81   *Karen Allen, a bio-psychologist at the University at Buffalo*   Allen, K. (2003). Are pets a healthy pleasure? The influence of pets on blood pressure. *Current Directions in Psychological Science,* 12(6), 236–239.

81   *Chinese women who owned dogs*   Headey, B., Na, F., & Zheng, R. (2008) Pet dogs benefit owner's health. A "natural experiment" in China. *Social Indicators Research*, 87(3), 481–493.

82   *an article in my local paper recently claimed*   Clark, P. (2010) Wagging tail, wet nose brighten hospital. *Asheville Citizen-Times.* The reference for the radiation therapy study is Johnson, R. A., Meadows, R. L., Haubner, J. S., & Sevedge, K. (2008). Animal-assisted activity among patients with cancer: Effects on mood, fatigue, self-perceived health, and sense of coherence. *Oncology Nursing Forum*, 35, 225–232.

82   *suffering from chronic fatigue syndrome*   Wells, D. L. (2009). Associations between pet ownership and self-reported health status in people suffering from chronic fatigue syndrome. *Journal of Alternative and Complementary Medicine*, 15, 407–413.

82   *effects of dog-walking on fitness*   Cutt, H. E., Knuiman, M. W., & Giles-Corti, B. (2008). Does getting a dog increase recreational walking? *International Journal of Behavioral Nutrition and Physical Activity,* 5(17).

83   *a study of 21,000 people in Finland*   Koivusilta, L. K., & Ojanlatva, A. (2006). To have or not to have a pet for better health? *PLoS ONE,* 1, 1–9.

83   *Researchers at the Australian National University*   Parslow, R. A., Jorm, A. F., Christensen, H., Rodgers, B., & Jacomb, P. (2005). Pet ownership and health in older adults: Findings from a survey of 2,551 community-based Australians aged 60–64. *Gerontology,* 51(1), 40–47.

83   *effects of acquiring a pet on loneliness in adults*   Gilbey, A., McNicholas, J., & Collis, G. M. (2007). A longitudinal test of the belief that companion animal ownership can help reduce loneliness. *Anthrozoös,* 20(4), 345–353.

83   *the results of thirty studies* Friedmann, E., & Son, H. (2009). The human-companion animal bond: How humans benefit. *Veterinary Clinics of North America*, 39(2) 293–326.

84   *Pet ownership does seem to make some, but not all, people feel healthier and happier* In anthrozoology, as in other areas of science, studies that show no effect are usually not published. This "file drawer effect" means that the published research probably exaggerates the actual effects of animals on human health. A recent study of twelve antidepressants found 33 of 36 experiments showing these drugs were ineffective were not published. Turner, E. H., Matthews, A. M., Linardatos, E., Tell, R. A., & Rosenthal, R. (2008). Selective publication of antidepressant trials and its influence on apparent efficacy. *New England Journal of Medicine*, 358(3), 252–260.

84   *There are three possibilities* McNicholas, J., Gilbey, A., Rennie, A., Ahmedzai, S., Dono, J. A., & Ormerod, E. (2005). Pet ownership and human health: A brief review of evidence and issues. *British Medical Journal*, 331(7527), 1252–1254.

84   *Karen Allen did it—with rich stockbrokers no less* Allen, K., Shykoff, B. E., & Izzo, J. L. (2001). Pet ownership, but not ACE inhibitor therapy, blunts home blood pressure responses to mental stress. *Hypertension*, 38(4), 815–820.

85   *sixteen seniors were brought to a Sydney emergency room* Kurrle, S. E., Day, R., & Cameron, I. D. (2004). The perils of pet ownership: A new fall-injury risk factor. *Medical Journal of Australia*, 181(11/12), 682–683.

85   *Centers for Disease Control* Centers for Disease Control (2009). Nonfatal fall-related injuries associated with dog and cats—United States, 2001–2006. *MMWR Weekly*, 58, 277–281.

85   *Pets can be health hazards in other ways as well* Some of the ways pets can be harmful are unusual. An article in the newsletter of the American Association of Reptile Veterinarians described eighteen cases of sexual aggression perpetuated by pet male iguanas toward their owners. Frye, F. F., Mader, D. R., & Centofanti, B. V. (1991). Interspecific (lizard:Human) sexual aggression in captive iguanas (*Iguana iguana*). *American Association of Reptile Veterinarians*, (1), 4–6.

85   *humans are susceptible to zoonotic* Pickering, L. K., Marano, N., Bocchini, J. A., & Angulo, F. J. (2008). Exposure to nontraditional pets at home and to animals in public settings: Risks to children. *Pediatrics*, 122(4), 876–886.

85   *pet-related Salmonella is on the rise* For a fascinating overview of diseases that humans can contract from animals, see Torrey, E. F., & Yolken, R. H. (2005). *Beasts of the earth: Animals, humans, and disease.* New Brunswick, NJ: Rutgers University Press.

85   *75,000 cases of Salmonella* Centers for Disease Control (2003) Reptile-Associated Salmonellosis—Selected States, 1998–2002. *MMWR*, 52(49) 1206–1209.

86   *spread MSRA* Lefebvre, S. L., & Weese, J. S. (2009). Contamination of pet

therapy dogs with MRSA and *Clostridium difficile*. *Journal of Hospital Infection*, 72, 268–269. Enoch, D., Karas, J., Slater, J., Emery, M., Kearns, A., & Farrington, M. (2005). MRSA carriage in a pet therapy dog. *Journal of Hospital Infection*, 60(2), 186–188.

86   *a wide variety of explanations for the human-animal bond* These theories are discussed in Tuan, Y. (1984). *Dominance affection: The making of pets*. New Haven, CT: Yale University Press. Grier, K. C. (2006). *Pets in America: A history*. Chapel Hill: University of North Carolina Press.; Franklin, A. (1999). *Animals and modern cultures: A sociology of human-animal relations in modernity*. London: Sage.; Serpell, J. (1996). *In the company of animals: A study of human-animal relationships*. Cambridge, UK: Cambridge University Press. Irvine, L. (2004). *If you tame me: Understanding our connection with animals*. Philadelphia, PA: Temple University Press. Olmert, M. D. (2009). *Made for each other: The biology of the human-animal bond*. Philadelphia, PA: Da Capo Press.

86   *Dan Gilbert of Harvard University claims that every psychologist* Gilbert, D. (2006). *Stumbling on happiness*. New York: Alfred A. Knopf.

87   *complex symbolic language, moral codes, religious beliefs, and the ability to learn to enjoy the burn of red hot chili peppers* For an excellent overview of the factors that make our species unique see Gazzaniga, M. (2008). *Human: The science behind what makes us unique*. New York: HarperCollins. For evidence that humans are the only mammal with the ability to learn to enjoy the burn of hot chilies, see Rozin, P., Gruss, L., & Berk, G. (1979). Reversal of innate aversions: Attempts to induce a preference for chili peppers in rats. *Journal of Comparative and Physiological Psychology*, 93(6), 1001–1014. There may be a few interesting exceptions. Rozin was able to teach two captive chimpanzees to like the taste of hot food. In addition, he found a few Mexican dogs that preferred piquant food. In all cases, these animals were raised as pets; Rozin believes that it was the social nature of the learning situation which enabled the animals to overcome natural taste aversions. Rozin, P., & Kenel, K. (1983). Acquired preferences for piquant foods by chimpanzees. *Appetite*, 4, 69–77.

87   *The pair were practically inseparable* You can watch the story of Tarra and Bella at www.elephants.com/tarra/TarraBella2.php.

87   *raising young rhesus monkeys with adult dogs* Mason, W. A., & Kenney, M. (1974). Redirection of filial attachments in rhesus monkeys: Dogs as mother surrogates. *Science*, 183(4130), 1209–1211.

87   *the chimpanzee* One incident has been used as evidence that chimps keep pets. The so-called pet was a hyrax, an animal that resembles a large guinea pig. In a scene reminiscent of *King Kong*, a team of Japanese primatologists working in Guinea observed a chimpanzee in a tree capture a hyrax, carefully carry it to the ground, and show it off to a couple of other chimps. Sounds like the start of a good relationship. But, just as in *King Kong*, things did not turn out well. The chimpanzee soon proceeded to smash the screaming hyrax against a tree trunk while his pals pummeled it with their fists until the little animal was

dead. After poking the hyrax's body around for a while, the apes got bored and casually tossed it into the bushes. The next morning, the researchers observed a female chimp grooming the hyrax's corpse for ten minutes. The chimpanzees may have treated the hyrax like toy, but they certainly did not think of it as a pet. This incident is reported in Hirata, S., Yamakoshi, G., Fujita, S., Ohashi, G., & Matsuzawa, T. (2001). Capturing and toying with hyraxes (*Dendrohyrax dorsalis*) by wild chimpanzees (*Pan troglodytes*) at Bossou, Guinea. *American Journal of Primatology*, 53(2), 93–97.

88 *best we can do is guess* There is a debate about whether modern modes of thinking evolved abruptly because of a sudden change in a handful of genes or occurred much more gradually. Balter, M. (2002) What made humans modern? *Science*, 295, 1219–1225; Wade, N. (July 25, 2006). Nice rats, nasty rats: Maybe it's all in the genes. *New York Times*. For discussions of the evolution of human cognitive abilities see Tomasello, M. (1999). *The cultural origins of human cognition*. Cambridge, MA: Harvard University Press; and Mithen, S. J. (1996). *The prehistory of the mind: The cognitive origins of art, religion and science*. London: Thames and Hudson. Serpell's views are in Serpell, J. A. (2003). Anthropomorphism and anthropomorphic selection-beyond the "cute response." *Society & Animals*, 11(1), 83–100.

89 *orgasm in human females* See, for example, Lloyd, E. A. (2005). *The case of the female orgasm: Bias in the science of evolution*. Cambridge, MA: Harvard University Press.

89 *attribute of human nature that evolved* See Serpell, J. (1996). *In the company of animals: A study of human-animal relationships*. Cambridge, UK: Cambridge University Press. (p. 148).

90 *human universals* Brown, D. E. (1991). *Human universals*. Philadelphia, PA: Temple University Press.; Pinker, S. (2002). *The blank slate: The modern denial of human nature*. New York: Viking.

90 *By comparing the two types of twins* Dunn, K. M., Cherkas, L. F., & Spector, T. D. (2005). Genetic influences on variation in female orgasmic function: A twin study. *Biology Letters*, 1(3), 260–263.; Silventoinen, K., Sammalisto, S., Perola, M., Boomsma, D. I., Cornes, B. K., Davis, C., et al. (2003). Heritability of adult body height: A comparative study of twin cohorts in eight countries. *Twin Research and Human Genetics*, 6(5), 399–408. Lykken, D., & Tellegen, A. (1996). Happiness is a stochastic phenomenon. *Psychological Science*, 7, 186–189.

91 *Steven Pinker's theory of music* Pinker, S. (1997). *How the mind works*. New York: Norton.

91 *nest parasitism, a reproductive strategy* The idea that pets are a kind of nest parasitism is discussed by Serpell (1996) and by Archer, J. (1997). Why do people love their pets? *Evolution and Human Behavior*, 18(4), 237–259

93 *Memes are everywhere* To learn more about cultural evolution and memes see Shennan, S. (2002). *Genes, memes and human history*. London: Thames & Hudson; and Blackmore, S. (1999). *The meme machine*. Oxford, UK: Oxford

University Press. The philosopher who has really taken Dawkins's idea and run with it is Daniel Dennett (1995). *Darwin's dangerous idea: Evolution and the meanings of life.* New York: Simon & Schuster. For a critical analysis of the meme idea see Richerson, P. J., & Boyd, R. (2005). *Not by genes alone: How culture transformed human evolution.* Chicago: University of Chicago Press.

93   *raised with pets usually grow up to be pet-owning adults* Anthropologists call the transmission of information from generation to generation "vertical" as compared to "horizontal transmission," which is the spread of memes across a culture.

94   *the pets of choice in Japanese homes* Bulliet, R. W. (2005). *Hunters, herders, and hamburgers: The past and future of human-animal relationships.* New York: Columbia University Press.

97   *Their smiles, wagging tails and kisses say it all.* Email from Ruby Benjamin (August 10, 2009).

97   *If dogs could talk* This statement was uttered by Dylan during his XM satellite radio show, Theme Time Radio Hour. The topic of the show was songs about dogs.

97   *genetic heritage runs 98% wolf and 2% dog* In reality, the claims about the "percent wolf" in a wolf-dog are almost never confirmed by actual genetic testing.

103   *Molecular biologists Heidi Parker and Elaine Ostrander* Parker, H. G., & Ostrander, E. A. (2005). Canine genomics and genetics: Running with the pack. *PLoS Genetics,* 1(5), e58.

103   *But when, where, and why did our ancestors* For overviews of recent findings on the evolution of dogs see Miklósi, A. (2007). *Dog behaviour, evolution, and cognition.* Oxford, UK: Oxford University Press.

103   *Paleolithic cave art is rife with stunning images* Guthrie, R. D. (2005). *The nature of Paleolithic art.* Chicago: University of Chicago Press. There was a tendency for early artists to concentrate on large mammals that they preyed upon. However, the lack of images of dogs does suggest that they did not have them as pets. Bulliet, R. W. (2005). *Hunters, herders, and hamburgers: The past and future of human-animal relationships.* New York: Columbia University Press.

104   *Domestication changes a species* Clutton-Brock, J. (1995). Origins of the dog: Domestication and early history. In J. Serpell (ed.), *The domestic dog: Its evolution, behaviour, and interactions with people* (pp. 7–20). Cambridge, UK: Cambridge University Press.

104   *The fossils suggest* The problem with old bones is that new ones turn up all the time, some of which don't fit conventional wisdom. For example, a group of archaeologists recently discovered fossil remains of canids that resemble large German shepherds in a Belgian cave that date back 32,000 years. Germonpré, M., Sablin, M. V., Stevens, R. E., Hedges, R. E. M., Hofreiter, M., Stiller, M., et al. (2009). Fossil dogs and wolves from Paleolithic sites in Belgium, the Ukraine, and Russia: Osteometry, ancient DNA and stable isotopes. *Journal*

of Archaeological Science, 36(2), 473–490. For the first evidence of a human-canine bond see Morey, D. F. (1994). The early evolution of the domestic dog. *American Scientist*, 82, 336–347.

105   *In 1997, an article appeared in the journal* Science Vila, C., Savolainen, P., Maldonado, J. E., Amorim, I. R., Rice, J. E., Honeycutt, R. L., et al. (1997). Multiple and ancient origins of the domestic dog. *Science*, 276, 1687–1689.

105   *Ray Coppinger, a biologist and dog-sled racer* Coppinger, R., & Coppinger, L. (2002). *Dogs: A new understanding of canine origin, behavior and evolution.* Chicago: University of Chicago Press.

106   *even hand-reared wolves* Topal, J., et al. (2010). The dog as a model for under-standing human social behavior. *Advances in the Study of Behavior.*

108   *an extraordinary experiment with foxes* Trut, L. N. (1999). Early canid domes-tication: The farm-fox experiment. *American Scientist*, 87, 160–169.

108   *a geneticist named Dmitri Belyaev* Belyaev and his brother supported Mendel's theory of genetics at a time when Lysenko's theory of inheritance was official Soviet policy. Belyaev's brother was sent to a concentration camp where he died. Wade, N. (July 25, 2006). Nice rats, nasty rats: Maybe it's all in the genes. *New York Times.*

108   *The foxes' physiology changed, too* Popova, N. K., Voitenko, N. N., Kulikov, A. V., & Avgustinovich, D. F. (1991). Evidence for the involvement of central se-rotonin in mechanism of domestication of silver foxes. *Pharmacology, Biochem-istry, and Behavior*, 40(4), 751–756. Lindberg, J., Björnerfeldt, S., Saetre, P., Svartberg, K., Seehuus, B., & Bakken, M. (2005). Selection for tameness has changed brain gene expression in silver foxes. *Current Biology*, 15(22), 915–916. It is less known that at the time he was starting the fox study, Belyaev also began similar experiments on the effects of selection for tameness on other animals including rats. After 70 generations of selective breeding, rats in the tamed strain loved to be petted and played with. Their mean brethren will rip your face off—even if the offspring of mean rats are raised by nice rat moms. Albert, F. W., Shchepina, O., Winter, C., Römpler, H., Teupser, D., & Palme, R. (2008). Phenotypic differences in behavior, physiology and neurochemistry between rats selected for tameness and for defensive aggression toward humans. *Hormones and Behavior*, 53(3), 413–421.

109   *"chicken-skin music."* I first heard this term from a 1976 Ry Cooder album of the same name.

109   *compared the ability of chimps, wolves, and dogs* Hare, B., & Tomasello, M. (2005). Human-like social skills in dogs? *Trends in Cognitive Sciences*, 9(9), 439–444.

109   *our closest, smartest, and most socially savvy relative* Some primatologists would argue that the bonobo, sometimes called the "pygmy chimpanzee," is just as smart or smarter.

109   *their brains are roughly 25% smaller.* Kruska, D. C. T. (2005). On the evolu-tionary significance of encephalization in some eutherian mammals: Effects of

adaptive radiation, domestication, and feralization. *Brain, Behavior and Evolution*, 65(2), 73–108.

110 *hot debate* For an objective overview of this argument see Morrell, V. (2009). Going to the dogs. *Science*, 325, 1062–1065.

110 *can train wolf pups* See Miklósi, A. (2007) and Gácsi, M. Györi, B., Virányi, Z., Kubiny, E., Range, F., Belényi, B. & Miklósi, A. (2009). Explaining dog wolf differences in utilizing human pointing gestures. *PLoS ONE*, 4(8), e6584.

110 *examples of canine brain power* Kaminski, J., Call, J., & Fischer, J. (2004). Word learning in a domestic dog: Evidence for "fast mapping." *Science*, 304(5677), 1682–1683. Rossi, A. P., & Ades, C. (2008). A dog at the keyboard: Using arbitrary signs to communicate requests. *Animal Cognition*, 11(2), 329–338. Range, F., Viranyi, Z., & Huber, L. (2007). Selective imitation in domestic dogs. *Current Biology*, 17(10), 868–872. Schwab, C., & Huber, L. (2006). Obey or not obey? Dogs (*Canis familiaris*) behave differently in response to attentional states of their owners. *Journal of Comparative Psychology*, 120(3), 169–175. Joly-Mascheroni, R., Senju, A., & Shepherd, A. J. (2008). Dogs catch human yawns. *Biology Letters*, 4, 446–448.

111 *They locate lost kids and rotting cadavers, warn deaf owners* For an overview of the many ways that we now use dogs to assist humans, see Zawistowski, S. (2008). *Companion animals in society*. Clinton Park, NY: Thompson.

111 *believe that their pets have a form of ESP* Sheldrake, R., & Smart, P. (2000). Testing a return-anticipating dog. *Anthrozoös*, 13(4), 203–212.

111 *a way to test what I call the Lassie Get Help hypothesis* Macpherson, K., & Roberts, W. A. (2006). Do dogs (*Canis familiaris*) seek help in an emergency? *Journal of Comparative Psychology*, 120(2), 113–119.

112 *Adam Miklósi's group* Gácsi, M., McGreevy, P., Kara, E., & Miklósi, A. (2009). Effects of selection for cooperation and attention in dogs. *Behavioral and Brain Function*, 5, 31.

112 *Researchers at the Center for the Interaction of Animals and Society* Hsu, Y., & Serpell, J. A. (2003). Development and validation of a questionnaire for measuring behavior and temperament traits in pet dogs. *Journal of the American Veterinary Medical Association*, 223(9), 1293–1300.

113 *trainability scores for 1,500 dogs of eleven breeds* Serpell, J. A., & Hsu, Y. (2005). Effects of breed, sex, and neuter status on trainability in dogs. *Anthrozoös*, 18(3), 196–207.

113 *Four and a half million Americans are bitten* Sacks, J. J., Kresnow, M., & Houston, B. (1996). Dog bites: How big a problem? *Injury Prevention*, 2(1), 52–54.

113 *bite strangers, turn on their owners, and pick fights with other dogs* Duffy, D. L., Hsu, Y., & Serpell, J. A. (2008). Breed differences in canine aggression. *Applied Animal Behaviour Science*, 114 (3–4), 441–460.

113 *Pit bulls were once considered a model of courage and loyalty* Other breeds have

been similarly stigmatized at different times including bloodhounds, German shepherds, Newfoundlands, and Dobermans. Delise, K. (2002). *Fatal dog attacks: The stories behind the statistic.* Manorville, NY: Anubis Press.

114 *the attacks did not stop* This account of the attack is based on a story that appeared in the September 10, 2008, *Seattle Times.*

115 *canine version of racial profiling* For the racial profiling argument, see Gladwell, M. (February 6, 2006). Troublemakers: What pit bulls can teach us about profiling. *The New Yorker.* Best Friends Animal Society is a major proponent of the campaign to rebrand pit bulls "America's Dog."

115 *Anthrozoologists are divided on breed bans* For an overview of the vicious breed issue, see the point-counterpoint exchange between Alan Beck and Ledy VanKakagage that appeared in the January 2007 *Veterinary Forum.*

115 *PETA wants pit bulls gone* In addition to supporting bans on pit bulls, PETA as well as the Humane Society of the United States took the position that the pit bulls confiscated from Michael Vick's dogfighting operation be euthanized because animals involved in fighting could not be rehabilitated. However, in April 2009, after considerable discussion, a group of organizations, including the ASPCA and HSUS, agreed to oppose blanket euthanasia of animals confiscated in raids on dogfights.

115 *owners of high-risk dogs* Barnes, J. E., Boat, B. W., Putnam, F. W., Dates, H. F., & Mahlman, A. R. (2006). Ownership of high-risk ("vicious") dogs as a marker for deviant behaviors: Implications for risk assessment. *Journal of Interpersonal Violence,* 21(12), 1616–1634. A more recent study also found that owners of "vicious breeds" had elevated histories of criminal behavior. Ragatz, L., Fremouw, W., Thomas, T., & McCoy, K. (2009). Vicious dogs: The antisocial behaviors and psychological characteristics of owners. *Journal of Forensic Sciences,* 54(3), 699–703.

116 *Between 1979 and 1990, rottweilers killed six people* Sacks, J. J., Kresnow, M., & Houston, B. (1996).

118 *article on the evolutionary psychology of human-animal relationships* Herzog, H. (2002). Darwinism and the study of human-animal interactions. *Society and Animals,* 10(4), 361–367.

118 *numbers of puppy registrations for each breed for the previous three years* Unfortunately, the AKC Web site no longer includes this information. However, annual registration statistics are published in the organization's magazine, *The AKC Gazette.*

119 *Their article was about names that people give their babies* Hahn, M. W., & Bentley, A. R. (2003). Drift as a mechanism for cultural change: An example from baby names. *Proceedings of the Royal Society B: Biological Sciences,* 270, 120–123.

119 *the 1,000 most common first names in the United States* Check the Baby Name Wizard Web site for a fascinating interactive graph showing how the popularity

of baby names rises and falls (www.babynamewizard.com/ voyager#prefix=& ms=true&sw=m&exact=false).

120   *compared power law graphs to a hockey stick* Gladwell, M. (February 13, 2006). Million dollar Murray. *The New Yorker*, pp. 96–107.

120   *"the long tail"* Anderson, C. (2008). *The long tail: Why the future of business is selling less of more.* New York: Hyperion.

120   *This left the other 125 breeds of dogs* The AKC typically admits a few new breeds each year. In 2009, three breeds were added, making the total 161.

121   *In 1962 they reached the tipping point* It is possible that they reached the tipping point because of the release of a Disney film called *Big Red* about this time that featured an Irish setter.

122   *a classic case of fashion trickle-down from the rich to the wanna-be-rich* Ritvo, H. (1987). *The animal estate: The English and other creatures in the Victorian age.* Cambridge, MA: Harvard University Press; Grier, K. C. (2006). *Pets in America: A history.* Chapel Hill: University of North Carolina Press.

123   *devastating critique in the* Atlantic Monthly See also Derr, M. (1990). The politics of dogs. *Atlantic Monthly,* 265(3), 49. See also Derr, M. (2004). *Dog's best friend: Annals of the dog-human relationship.* Chicago: University of Chicago Press.

124   *purebred dogs have become a favorite animal model for the study of human diseases* For an excellent overview of current research showing how studies of dog genetics have been useful in understanding human diseases, see Ostrander, E. A., & Wayne, R. K. (2005). The canine genome. *Genome Research,* 15(12), 1706–1716.

125   *trace their ancestry to thirty-one animals* Chase, K., Sargan, D., Miller, K., Ostrander, E., & Lark, K. (2006). Understanding the genetics of autoimmune disease: Two loci that regulate late onset Addison's disease in Portuguese water dogs. *International Journal of Immunogenetics,* 33(3), 179–184.

126   *as many as 90% of adoptable dogs in some parts of the Northeast* Jonsson, P. (June 18, 2008). "Dixie dogs" head north. *Christian Science Monitor.* Herzog, H. (January 26, 2009) A first dog from down South. *Washington Post.* Peters, S. L. (April 23, 2009) Doomed dogs get on rescue wagon to other shelters. *USAToday.*

129   *As a society, we expect women* Luke, B. (2007). *Brutal: Manhood and the exploitation of animals.* Urbana: University of Illinois Press. (p. 15).

129   *Men enjoy hunting and killing* Washburn, S. L, & Lancaster, C. S. (1968). The evolution of hunting. In R. B. Lee & I. DeVore (eds.) *Man the hunter.* Chicago: Aldine Publishing. (p. 299).

129   *interviewed veterinary students* This study is reported in Herzog, H. A., Vore, T. L., & New, J. C. (1989). Conversations with veterinary students: Attitudes, ethics, and animals. *Anthrozoös, 2,* 181–188.

132   *amassing every study* For a summary of these findings, see Herzog, H. A.

(2006). Gender differences in human-animal interactions. *Anthrozoös*, 20(1), 7–21.

132 *equal numbers of men and women own companion animals* Pew Research Center (March 7, 2006). Gauging family intimacy: Dogs edge cats (dads trail both). *Pew Research Center: A Social Trends Report.* This study, which was based on a random sample of 3,000 American adults, found that 56% of men and 57% of women owned a pet. Interestingly, equal numbers of men and women owned dogs (40% versus 39%) and cats (21% versus 24%). Women were slightly more likely than men to consider their pets members of their family. The survey also found that the subjects felt closer to their dogs and cats than they did to their parents.

132 *pay newspapers to publish obituaries* Wilson, C., Netting, F., Turner, D., Roth, C., & Olsen (2009) Companion animal obituaries: The "hairy heirs." Presentation to the International Society for Anthrozoology, Kansas City.

132 *"My pet means more to me than any of my friends."* These items are from the Lexington Attachment to Pets Scale. Johnson, T., Garrity, T., & Stallones, L. (1992). Psychometric evaluation of the Lexington Attachment to Pets Scale (LAPS). *Anthrozoös*, 5(3), 160–175.

133 *play chase and pull-and-tug with them* You might not want to follow our example. Some animal behaviorists believe that playing tug-of-war with dogs gives them the message that it is OK for them to kick your butt and creates long-term dominance issues. The evidence for this view is mixed.

133 *how they played with their dogs* Prato-Previde, E., Fallani, G., & Valsecchi, P. (2006). Gender differences in owners interacting with pet dogs: An observational study. *Ethology*, 112(1), 64–73.

133 *their owners in a veterinary waiting room* Mallon, G. (1993). A study of the interactions between men, women, and dogs at the ASPCA in New York City. *Anthrozoös*, 6, 43–47.

133 *Gail Melson, a developmental psychologist* Melson, G. F., & Fogel, A. (1996). Parental perceptions of their children's involvement with house-hold pets. *Anthrozoös*, 9, 95–106. For an excellent overview of the roles that animals play in the development of children, see Melson, L. G. (2001). *Why the wild things are: Animals in the lives of children.* Cambridge, MA: Harvard University Press.

133 *creatures that are cute* Sprengelmeyer, R., Perrett, D., Fagan, E., Cornwell, R., Lobmaier, J., Sprengelmeyer, A., et al. (2009). The cutest little baby face: A hormonal link to sensitivity to cuteness in infant faces. *Psychological Science*, 20(2), 149–154; Fridlund, A., & MacDonald, M. (1998). Approaches to Goldie: A field study of human approach responses to canine juvenescence. *Anthrozoös*, 11, 95–100.

134 *Stephen Kellert of Yale University* Kellert, S. R. (1996). *The value of life: Biological diversity and human society.* Washington, DC: Island Press.; Kellert, S. R., & Berry, J. K. (1987). Attitudes, knowledge, and behaviors toward wildlife as affected by gender. *Wildlife Society Bulletin*, 15(3), 363–371.

134   *phobias of creatures like snakes and spiders* Fredrikson, M., Annas, P., Fischer, H. Å., & Wik, G. (1996). Gender and age differences in the prevalence of specific fears and phobias. *Behaviour Research and Therapy,* 34(1), 33–39.

134   *felt about using dogs and chimpanzees in experiments* Pifer, L., Shimizu, K., & Pifer, R. (1994). Public attitudes toward animal research: Some international comparisons. *Society and Animals,* 2(2), 95–113.

134   *Swedish researchers* Hagelin, J., Carlsson, H. E., & Hau, J. (2003). An overview of surveys on how people view animal experimentation: Some factors that may influence the outcome. *Public Understanding of Science,* 12(1), 67.

134   *the National Opinion Research Center* This data is from the 1993 General Social Survey.

136   *Victorian era animal rights crusaders* French, R. D. (1975). *Antivivisection and medical science in Victorian society.* Princeton, NJ: Princeton University Press.

136   *Women dominate nearly every aspect of grassroots animal protection* While women make up the bulk of grassroots animal activists, this is not true of the leadership of the animal rights movement. For example, women are underrepresented among lists and biographical indexes of the movers and shakers of the animal protection movement. See Herzog, H. (November 1999) Power, money and gender: Status hierarchies and the animal protection movement in the United States. *International Society of Anthrozoology Newsletter,* 2–5.

137   *that 70% of the battered women* Ascione, F. (1998). Battered women's reports of their partners' and their children's cruelty to animals. *Journal of Emotional Abuse,* 1(1), 119–133. Similar results were reported in a study of battered women in South Carolina; see Flynn, C. P. (2000). Woman's best friend: Pet abuse and the role of companion animals in the lives of battered women. *Violence against Women,* 6(2), 162–177. For a recent review, see Ascione, F. R., et al. (2007). Battered pets and domestic violence: Animal abuse reported by women experiencing intimate violence and by non-abused women. *Violence against Women,* 13, 354–373.

137   *The data at Pet-abuse.com* The person who realized that Pet-abuse.com offered a window into the demographics of animal cruelty was Kathy Gerbasi, a developmental psychologist; see Gerbasi, K. C. (2004). Gender and nonhuman animal cruelty convictions: Data from www.pet-abuse.com. *Society and Animals,* 12(4), 359–365. These statistics were downloaded from www.pet-abuse.com in January 2009.

139   *"The Great Bunny Rescue of 2006"* Weise, E. (August 8, 2006). Rabbit rescue ends some bad hare days. *USAToday.*

139   *public health implication of animal hoarding* Hoarding of Animals Research Consortium (2002). Health implications of animal hoarding. *Health & Social Work,* 27(2), 125–132.

139   *explanations of why people hoard animals* Patronek, G., Loar, L., & Nathanson, J. N. (2006). *Animal hoarding: Structuring interdisciplinary responses to*

help people, animals, and communities at risk. Boston: Hoarding of Animals Research Consortium.

139   *caused by toxoplasmosis infection* Sklott, R. (December 9, 2007). "Cat Lady" conundrum. *New York Times.*

139   *animal hoarding is caused by dementia* The most common theory is that animal hoarding is a type of obsessive compulsive disorder. However, the validity of this view has recently been questioned; see Patronek, G. (2007). Animal hoarding: What caseworkers need to know. Paper presented at the MassHousing Community Services Conference, Boston, MA.

140   *The recidivism rate among hoarders* Miller, S. (January/February 2008). Objects of their affection: The hidden world of hoarders. *Best Friends Magazine,* 20–22, 57–61. Non-animal hoarding has been linked to damage to the prefrontal cortex of the brain; see Anderson, S. W., Damasio, H., & Damasio, A. R. (2005). A neural basis for collecting behaviour in humans. *Brain,* 128(1), 201–212. No one has tested the possibility that animal hoarding can be caused by brain damage.

141   *other director is in the wrong line of work* For an ethnographic study of how animal shelter workers cope with the paradox of loving animals and killing them see Arluke, A. (2006). *Just a dog: Understanding animal cruelty and ourselves.* Philadelphia, PA: Temple University Press.

142   *some social scientists* See, for example, Adams, C. J. (2002). *The sexual politics of meat: A feminist-vegetarian critical theory.* New York: Continuum Publishing. Donovan, J., & Adams, C. J. (eds.) (2007). *The feminist care tradition in animal ethics.* New York: Columbia University Press.

142   *almost every human culture* The examples of women hunters in tribal society are from Estioko-Griffin, A., & Griffin, P. B. (1981). Woman the hunter: The Agta. In F. Dahlbrg (ed.), *Woman the gatherer* (pp. 121–140). New Haven, CT: Yale University Press. Bailey, R. C., & Aunger, R. (1989). Net hunters vs. archers: Variation in women's subsistence strategies in the Ituri forest. *Human Ecology,* 17(3), 273–297.; Romanoff, S. (1983). Women as hunters among the Matses of the Peruvian amazon. *Human Ecology, 11*(3), 339–343; Goodman, M. J., Griffin, P. B., Estioko-Griffin, A. A., & Grove, J. S. (1985). The compatibility of hunting and mothering among the Agta hunter-gatherers of the Philippines. *Sex Roles,* 12(11), 1199–1209.

143   *In nearly all human societies* Wood, W., & Eagly, A. H. (2002). A cross-cultural analysis of the behavior of women and men: Implications for the origins of sex differences. *Psychological Bulletin,* 128(5), 699–727.

143   *the result of socialization* Quinn, P. C., & Liben, L. S. (2008). A sex difference in mental rotation in young infants. *Psychological Science,* 19(11), 1067–1070. Moore, D. S., & Johnson, S. P. (2008). Mental rotation in infants. *Psychological Science,* 19(11), 1063–1066. Williams, C. L., & Pleil, K. E. (2008). Toy story: Why do monkey and human males prefer trucks? Comment on "Sex differences in rhesus monkey toy preferences parallel those of children" by Hassett, Siebert, and Wallen. *Hormones and Behavior,* 54, 355–358. Rakison, D. H. (2009). Does

women's greater fear of snakes and spiders originate in infancy? *Evolution and Human Behavior.* 30, 438–444.

143   *empathy* Taylor, N., & Signal, T. (2005). Empathy and attitudes to animals. *Anthrozoös.* 18(1), 18–27. For a review of the literature on the connection between love of pets and love of people, see Paul, E. S. (2000). Love of pets and love of people. In A. Podberscek, E. S. Paul & J. A. & Serpell (eds.), *Companion animals and us: Exploring the relationships between people and animals.* (pp. 168–186). Cambridge, UK: Cambridge University Press.

143   *makes men more generous with their money* Domes, G., Heinrichs, M., Michel, A., Berger, C., & Herpertz, S. C. (2007). Oxytocin improves "Mind-reading" in humans. *Biological Psychiatry,* 61(6), 731–733.; Zak, P. J., Stanton, A. A., & Ahmadi, S. (2007). Oxytocin increases generosity in humans. *PLoS ONE,* 2(11), e1128.

143   *Is oxytocin the glue* Odendaal, J. S. J., & Meintjes, R. (2003). Neurophysiological correlates of affiliative behaviour between humans and dogs. *The Veterinary Journal,* 165(3), 296–301. Miller, S. C., Kennedy, C., DeVoe, D., Hickey, M., Nelson, T., & Kogan, L. (2009). An examination of changes in oxytocin levels in men and women before and after interaction with a bonded dog. *Anthrozoös,* 22(1), 31–42. Nagasawa, M., Kikusui, T., Onaka, T., & Ohta, M. (2009). Dog's gaze at its owner increases owner's urinary oxytocin during social interaction. *Hormones and Behavior,* 55(3), 434–441. The study showing no effect of oxytocin, as is often the case with negative results, has not been published even though there were plenty of subjects and the lead researcher is a well-known anthrozoologist. (Personal communication, Rebecca Johnson, January 3, 2010.)

144   *more aggressive and the less empathetic you are* Harris, J. A., Rushton, J. P., Hampson, E., & Jackson, D. N. (1996). Salivary testosterone and self-report aggressive and pro-social personality characteristics in men and women. *Aggressive Behavior,* 22(5), 321–331.

145   *Height is a good example* I first realized the implications of mathematics of overlapping bell curves after reading Gladwell's *New Yorker* article. However, the example using height is taken from Steven Pinker's excellent discussion of the bell curve in *The Blank Slate.*

146   *anorexia* Hoek, H. W., & van Hoeken, D. (2003). Review of the prevalence and incidence of eating disorders. *International Journal of Eating Disorders,* 34(4), 383–396.

149   *Cleveland Amory* In Dundes, A. (1994). *The cockfight: A casebook.* Madison: University of Wisconsin Press. (p. 71).

149   *Captain L. Fitz-Barnard* Fitz-Barnard, L. (1921). *Fighting sports.* London: Oldam Press. (p. 12).

153   *a fifty-year-old black man* In the late 1970s, the rooster fights were probably the most integrated Saturday night activity in Madison County. While most of the spectators and participants were white, Doc was the highest-status referee, and it was common to find a few blacks mingling among the spectators and

rooting for the chickens they had money on. At one point during my research, a couple of outside organizers for the Ku Klux Klan tried to stage a rally in the county. They initially approached the chair of the county school board about having the rally at the high school, but were rebuffed. The owner of the cock pit was more accommodating. The outsiders dressed up in their silly white robes and held their meeting at the pit. Hardly anyone showed up except the documentary photographer Rob Amberg, and the racists left town the next day, having accomplished nothing. However, no one bothered to take down the KKK rally announcements that were posted around the pit. When I attended a derby a few weeks later, Doc was still the referee and the black guys from Asheville were placing bets just like everyone else. No one but me seemed to notice the KKK signs. This easy mixing of races was similarly characteristic of cockfights in Louisiana. See Maunula, M. (2007). Of chickens and men: Cockfighting in the South. *Southern Cultures*, 13(4), 76–85.

154 *Gregory Bateson and Margaret Mead wrote* For a compendium of historical and anthropological writings on the significance of cockfighting in different cultures see Dundes, A. (1994). *The cockfight: A casebook*. Madison: University of Wisconsin Press. In addition, there are several fictional treatments of cockfighting and cockfighting culture. These include West, N. (1995). *Day of the locust*. New York: Bantam Books. Willeford, C. E. (1987). *Cockfighter*. New York: Creative Arts Books. Manley, F. (1998). *The cockfighter*. Minneapolis: Coffee House Press. For essays on cockfighting as cultural phenomena see Bilger, B. (2000). *Noodling for flatheads: Moonshine, monster catfish, and other Southern comforts*. New York: Touchstone; and Crews, H. (1977). Cockfighting: An unfashionable view. *Esquire*, 87, 8, 12, 14.

154 *"homoerotic male battle with masturbatory nuances."* Dundes, A. (1994). The gallus as phallus. In A. Dundes, *The cockfight: A casebook*. (pp. 241–281). Madison: University of Wisconsin Press. (p. 262.)

155 *oldest and most widespread of traditional sports* For a history of cockfighting see Smith, P., & Daniel, C. (2000). *The chicken book*. Athens: University of Georgia Press.

157 *the basic set of guidelines, referred to as Wortham's Rules* Wortham's Rules apply to the gaff fights that were preferred by Appalachian cockfighters. Hispanic cockfighters typically use a different set of rules.

157 *a Platonic concept of hate."* Geertz, C. (1994). Deep play: Notes on the Balinese cockfight. In A. Dundes (ed), *The cockfight: A casebook*. (p. 103). Madison: University of Wisconsin Press.

161 *the sociologist Clifton Bryant estimated* Bryant, C. (1991). Deviant leisure and clandestine lifestyle: Cockfighting as a socially disvalued sport. *World Leisure and Recreation*, 33(2), 17–21.

163 *Captain L. Fitz-Barnard writes* Fitz-Barnard, L. (1921). *Fighting sports*. London: Oldam Press. (p. 12).

163 *120 million wild birds shot by hunters* Klem, D. (1991). Glass and bird kills. An overview and suggested planning and design methods for preventing a fatal

hazard. In L. W. Adams & D. L. Leedy (eds.). *Wildlife conservation in metropolitan environments*. (pp. 99–103). Columbia, MD: National Institute for Urban Wildlife.

163 *most cockfighters I met did not approve of dogfighting* Suzie, the Louisiana animal protectionist, agrees with Eddy's wife that the cultures of cockfighting and of dogfighting are different. In her experience, you are much more likely to find pornography, drugs, and large sums of money when raiding the homes of urban dogfighters than rural cockfighters.

164 *the Humane Society of the United States links cockfighting with* www.hsus.org/acf/fighting/cockfight/ cockfighting_and_related_crimes.html. Some of these charges are true. Illegal gambling (and hence tax evasion) is part of virtually every cockfight. I do not doubt that now, as was the case back in the 1970s, some local officials are paid by pit owners to turn a blind eye, and that the influx of Hispanic cockfighters has meant more illegal immigrants show up at the pits.

165 *That's what's kept me interested in it.* The quote is by Bobby Keener of Greensboro, North Carolina. It is from the DVD *Cockfighters,* an eight-hour series of interviews with cockfighters produced by Olena Media (www.olenamedia.com/).

165 *he wrote in* Grit and Steel McCaghy, C. H., & Neal, A. G. (1974). The fraternity of cockfighters: Ethical embellishments of an illegal sport. *Journal of Popular Culture*, 8(3), 557–569. (p. 74).

165 *Cocks are not usually fought until they are two years old* Younger cocks are sometimes fought in matches called stag fights.

167 *Karen Davis, the founder of United Poultry Concerns* For Karen's perspective on the treatment of chickens, see Davis, K. (2009). *Prisoned chickens, poisoned eggs: An inside look at the modern poultry industry.* Summertown, TN: Book Publishing Company.

167 *two corporate giants, Tyson Foods and Upjohn* In 1994 Tyson bought out 100% of the Upjohn stock.

169 *their trip to the processing plant* For an exposé of the treatment of chickens and other animals in American slaughterhouses, see Eisnitz, G. A. (2007). *Slaughterhouse: The shocking story of greed, neglect, and inhumane treatment inside the U.S. meat industry.* Amherst, NY: Prometheus Books.

170 *the neck cutting machine* Controlled atmosphere killing (CAK) is an alternative to electrical stunning and death by neck-cutting machine. CAK is essentially a gas chamber in which birds are exposed either to a stunning or lethal dose of carbon dioxide or an inert gas such as argon or nitrogen. The Humane Society of the United States recommends CAK as a more humane alternative to electrical stunning. The National Chicken Council disagrees. See Shields, S., & Raj, M. (no date). *An HSUS report. The welfare of birds at slaughter.* The Humane Society of the United States. See also Savage, C. (February 17, 2009). No advantage to gas-based stunning for chickens. *Meat & Poultry.*

170 *An efficient production system with two evisceration lines will process 140 birds per minute* Bell, D. D., & Weaver, W. D. (2001). *Commercial chicken meat*

and egg production. New York: Springer (p. 903). Even when birds are killed by hand, as in the case of Kosher slaughter, the rate of production is astounding. Wabeck indicates that one worker should be able to slit the throats of 4,000 birds per hour in Bell, D. D., & Weaver, W. D. (2001).

170 *Some birds are not completely stunned* Karen Davis of United Poultry International argues that the stunning procedure is ineffective and was not developed to reduce suffering. In an email to me, she wrote, "The purposes of administering electricity to the bodies of birds in the slaughterhouse is not electrocution—to kill them outright—but merely muscle paralysis to facilitate feather removal and keep the birds still on the conveyer belt. Stunning was never designed to be a 'humane' method. It was designed in the 1930s for strictly commercial purposes—a technological 'upgrade' from paralyzing birds through the roof of their mouths to their brains with a knife or other sharp object." Her view is supported by Charles Wabeck who wrote the chapter on broiler processing in Bell, D. D., & Weaver, W. D. (2001). *Commercial chicken meat and egg production*. New York: Springer. Wabeck writes, "Stunning is essential for satisfactory bleeding and feather release" (p. 904). Nowhere in his section on killing is the issue of animal welfare discussed.

172 *cockfighting became illegal in every state* Cockfighting, however, remains legal in the American territories of Puerto Rico, Guam, Samoa, and the Virgin Islands.

172 *The eighteenth-century movement against blood sports* See Thomas, K. (1983). *Man and the natural world: A history of the modern sensibility.* New York: Pantheon.

172 *essay that appeared in the* Atlanta Journal Constitution Rudy, K. (September 6, 2007). Dog-fighting and Michael Vick. *Atlanta Journal Constitution.*

172 *You see the same thing in thoroughbred racing* McMurray, J. (June 14, 2008). AP finds 5K horse deaths since '03. Washingtonpost.com.

175 *You ask me why I refuse to eat flesh* Coetzee, J. M. (2003). *Elizabeth Costello.* New York: Penguin Books.

176 *sixfold in Japan and fifteenfold in China* Halweil, B., & Nierenberg, D. (2008). Meat and seafood: The global diet's most costly ingredients. In Worldwatch Institute (ed.), *2008 State of the World: Innovation for a sustainable economy* (pp. 61–74). Washington, DC: Worldwatch Institute.

176 *less about your cholesterol than your sanity* Bruni, F. (May 20, 2009). Beef and décor, aged to perfection. *New York Times.*

177 *Chimpanzees, our closest living relatives, love the taste of meat* For more information on chimpanzee carnivory, see Stanford, C. B. (1999). *The hunting ape: Meat eating and the origins of human behavior.* Princeton, NJ: Princeton University Press. Stanford, C. (2002). *Significant others: The ape-human continuum and the quest for human nature.* New York: Basic Books. Boesch, C. (1994). Hunting strategies of Gombe and Taï chimpanzees. In R. W. Wrangham, W. C. McGrew, F. B. M. de Waal, & P. G. Heltne (eds.), *Chimpanzee cul-*

tures (pp. 76–92). Cambridge, MA: Harvard University Press. Foley, R. (2001). The evolutionary consequences of increased carnivory. In C. B. Stanford & H.T. Bunn (eds)., *Meat-eating and human evolution* (pp. 305–331). New York: Oxford University Press. The exchange of meat for sex is described in Gomes, C. M., & Boesch, C. (2009). Wild chimpanzees exchange meat for sex on a long-term basis. *PLoS ONE*, 4, e5116.

178 *wrote that our ancestors were blood-thirsty killers* Stanford, C. B. (1999). *The hunting ape: Meat eating and the origins of human behavior.* Princeton, NJ: Princeton University Press. (p. 107).

179 *the caloric intake of the Nunamuit people of northern Alaska* Cordain, L., Eaton, S., Brand Miller, J., Mann, N., & Hill, K. (2002). The paradoxical nature of hunter-gatherer diets: Meat-based, yet non-atherogenic. *European Journal of Clinical Nutrition*, 56(1), 42–52.; Gadsby, P. (October 1, 2004). The Inuit paradox. *Discover Magazine*.

179 *Not a single hunter-gatherer society* Cordain, L., Eaton, S. B., Sebastian, A., Mann, N., Lindeberg, S., Watkins, B. A., et al. (2005). Origins and evolution of the Western diet: Health implications for the 21st century. *American Journal of Clinical Nutrition*, 81(2), 341–354.

181 *susceptible to the various bacteria, viruses, protozoans* Torrey, E. F. (2005). *Beasts of the earth: Animals, humans, and disease.* New Brunswick, NJ: Rutgers University Press; Finch, C. E., & Stanford, C. B. (2004). Meat-adaptive genes and the evolution of slower aging in humans. *Quarterly Review of Biology*, 79(1), 3–50. Chitnis, A., Rauls, D., & Moore, J. (2000) Origin of HIV Type 1 in colonial French Equatorial Africa? *AIDS Research and Human Retroviruses*, 16(1) 5–8.

181 *humans are unique among animals in spicing food* Hot chilies are also aversive to human children. Studies by Paul Rozin have shown that humans have to learn to enjoy the burn of chilies. Rozin, P., & Schiller, D. (1980). The nature and acquisition of a preference for chili pepper by humans. *Motivation and Emotion*, 4(1), 77–101.

182 *eating flesh is still risky* Fessler, D. M. T. (2002). Reproductive immunosuppression and diet. *Current Anthropology*, 43(1), 19–61. Flaxman, S. M., & Sherman, P. W. (2000). Morning sickness: A mechanism for protecting mother and embryo. *Quarterly Review of Biology*, 75(2), 113–148.

184 *aversions to meat are three times more common than aversions to vegetables and six times more common than fruit aversions* Midkiff, E. E., & Bernstein, I. L. (1985). Targets of learned food aversions in humans. *Physiology & Behavior*, 34(5), 839–841.

184 *food taboos across human societies* Fessler, D., & Navarrete, C. (2003). Meat is good to taboo: Dietary proscriptions as a product of the interaction of psychological mechanisms and social processes. *Journal of Cognition and Culture*, 3(1), 1–40.

185 *a taboo against eating buffalo meat* This is described by McDonaugh, C.

(1997). Breaking the rules: Changes in food acceptability among the Tharu of Nepal. In H. Macbeth (ed.), *Food preferences and taste: Continuity and change*. Oxford, UK: Berghahn Books.

185 *For most Americans, the idea of eating dogmeat* Not all Americans are turned off by dogmeat. For Filipinos living in California, dogmeat is a traditional food served at important ceremonies such as weddings. The practice of eating dogs has caused conflict between Filipinos and Anglos in California. In response to a 1989 incident in which a group of Cambodian refugees killed and skinned a German shepherd puppy to eat, the California legislature enacted Penal Code section 598b, which banned the possession, sale, import, or giving away the carcass of "any animal traditionally or commonly kept as a pet" for use as food. See Griffith, M., Wolch, J., Lassiter, U. (2002). Animal practices and the racialization of Filipinas in Los Angeles. *Society & Animals*, 10(3), 221–248.

185 *humans have been eating dogs for thousands of years* Discussions of historical and cultural variation in attitudes toward eating dog are found in McHugh, S. (2004). *Dog*. London: Reaktion Books. Serpell, J. (1995). The hair of the dog. In J. Serpell (ed.), *The domestic dog: Its evolution, behavior and interactions with people* (pp. 257–262). Cambridge, UK: Cambridge University Press; Simoons, F. J. (1994).

186 *the Asian trade in dog products* For an overview of the status of dog-eating in Asia see Podberscek, A. (2009). Good to pet and eat: The keeping and consuming of dogs and cats in South Korea. *Journal of Social Issues*, 65, 615–632. Walraven, B. (2002). Bardot soup and Confucians' meat: Food and Korean identity in global context. In K. Cwiertka & B. Walraven (eds.), *Asian food: The global and the local* (pp. 95–115). Honolulu: University of Hawaii Press.

187 *In classical Hinduism* Nelson, L. (2006). Cows, elephants, dogs, and other lesser embodiments of Atman: Reflections of Hindu attitudes toward nonhuman animals. In P. Waldau & K. Patton, *A communion of subjects: Animals in religion, science, and ethics* (pp. 179–193). New York: Columbia University Press.

187 *Islamic law also regards dogs as unclean* Foltz, R. (2006). "This she-camel of God is a sign to you": Dimensions of animals in Islamic tradition and Muslim culture. In P. Waldau & K. Patton (eds.), *A communion of subjects: animals in religion, science, and ethics* (pp. 149–150). New York: Columbia University Press.

187 *Oglala Indians* Powers, W., & Powers, M. (1986). Putting on the dog. *Natural History*, 2, 6–16.

188 *Ceremonies in which hunters* Serpell, J. (1996). *In the company of animals: A study of human-animal relationships*. Cambridge, UK: Cambridge University Press.

188 *animal liberation philosopher Peter Singer would have few objections to* Singer, P., & Mason, J. (2006). *The ethics of what we eat: Why our food choices matter*. Emmaus, PA: Rodale Press.

189 *Sandy McGee and I distributed questionnaires* Herzog, H., & McGee, S. (1983). Psychological aspects of slaughter: Reactions of college students of killing and butchering cattle and hogs. *International Journal for the Study of Animal Problems*, 4(2), 124–132.

189 *He writes, "Humans must eat, excrete, and have sex"* Rozin, P., Haidt, J., & McCauley, C. R. (2000). Disgust. In M. Lewis & J. M. Haviland-Jones (eds.), *Handbook of emotions* (2nd ed.) (p. 642). New York: Guilford Press.

190 *researchers* Kubberød, E., Ueland, Ø., Dingstad, G. I., Risvik, E., & Henjesand, I. J. (2008). The effect of animality in the consumption experience: A potential for disgust. *Journal of Food Products Marketing*, 14(3), 103–124; and Kubberød, E., Ueland, Ø., Rødbotten, M., Westad, F., & Risvik, E. (2002). Gender specific preferences and attitudes toward meat. *Food Quality and Preference*, 13(5), 285–294.

190 *moralization* Rozin, P., Markwith, M., & Stoess, C. (1997). Moralization and becoming a vegetarian: The transformation of preferences into values and the recruitment of disgust, *Psychological Science*, 8, 67–73.

191 *The case against meat boils down to four claims* The literature on the ethical, ecological, health and feminist arguments against meat is vast. For a start, I suggest reading Singer, P., & Mason, J. (2006). *The ethics of what we eat: Why our food choices matter.* Emmaus, PA: Rodale Press; Eisnitz, G. A. (2007). *Slaughterhouse: The shocking story of greed, neglect, and inhumane treatment inside the U.S. meat industry.* Amherst, NY: Prometheus Books; and Adams, C. J. (2000). *The sexual politics of meat: A feminist-vegetarian critical theory.* London: Continuum International Publishing Group. The 2009 film *Food, Inc.* offers a graphic depiction of the treatment of animals and human workers on modern factory farms.

191 *the Vegetarian Resource Group* The results of the Vegetarian Resource Group surveys can be found at www.vrg.org/nutshell/faq.htm#poll.

192 *Food and Drug Administration told us to cut down on saturated fats* The history of the rise of "nutritionism" in the United States is described in Pollan, M. (2008). *In defense of food: An eater's manifesto.* New York: Penguin Press.

193 *the average American ate a half pound of chicken a year* Boyd, W., & Watts, M. (1997). Agro-industrial just-in-time: The chicken industry and postwar American capitalism. In D. Goodman & M. Watts (eds.), *Globalizing food: Agrarian questions and global restructuring* (pp. 192–225). New York: Routledge.

193 *early claims against the hazards of red meat were based on shoddy science* Pollan, M. (2008).

193 *multisite study of half a million people* Sinha, R., Cross, A. J., Graubard, B. I., Leitzmann, M. F., & Schatzkin, A. (2009). Meat intake and mortality: A prospective study of over half a million people. *Archives of Internal Medicine*, 169, 562–571.

194 *how Newkirk explained PETA's logic to me* Ingrid Newkirk (email) June 24, 2009.

194 *Humane Research Council* The Humane Research Council also serves as a clearing house for research on human-animal interactions. Their Web site is one of the best sources of information about the latest studies in anthrozoology: www.humaneresearch.org/.

195 *in 2002,* Time *reported* Corliss, R. (July 15, 2002). Should we all be vegetarians? *Time* magazine.

195 *One tenth of one percent of Americans are strict vegetarians* The USDA results are found in Fields, C., Dourson, M., & Borak, J. (2005). Iodine-deficient vegetarians: A hypothetical perchlorate-susceptible population? *Regulatory Toxicology and Pharmacology,* 42:37–46.

197 *But they were also more likely to be anxious and worried* The vegetarians scored higher than meat-eaters on the Big Five traits of openness to experience and neuroticism (traits I also score high on). Golden, L., & Herzog, H. (2008). Presentation to the meeting of the Southeastern Psychological Association, New Orleans. For a description of the Five Factor Model of Personality, see Gozling, S. (2008). *Snoop: What your stuff says about you.* New York: Basic Books.

198 *vegetarianism is most common among teenage girls* The 2007 Harris poll found that 10% of girls between the ages of thirteen and eighteen said they never ate meat or seafood. Moskin, J. (January 24, 2007). Strict vegan ethics, frosted with hedonism. *New York Times.*

198 *research on the connection between vegetarianism and eating disorders* Research linking vegetarianism and eating disorders can be found in Klopp, S. A., Heiss, C. J., & Smith, H. S. (2003). Self-reported vegetarianism may be a marker for college women at risk for disordered eating. *Journal of the American Dietetic Association,* 103(6), 745–747. Neumark-Sztainer, D., Story, M., Resnick, M., & Blum, R. (1997). Adolescent vegetarians. A behavioral profile of a school-based population in Minnesota. *Archives of Pediatrics and Adolescent Medicine,* 151(8), 833–838. O'Connor, M. A., Touyz, S. W., Dunn, S. M., & Beumont, P. J. (1987). Vegetarianism in anorexia nervosa? A review of 116 consecutive cases. *The Medical Journal of Australia,* 147(11–12), 540–542. Robinson-O'Brian, R., Perry, C. L.,Wall, M. M., Story, M., & Neumark-Sztainer, D. (2009). Adolescent and young adult vegetarianism: Better dietary intake and weight outcomes but increased risk of disordered eating behaviors. *Journal of the American Dietetic Association,* 109, 648–655. Worsley, A., & Skrzypiec, G. (1997). Teenage vegetarianism: Beauty or the beast? *Nutrition Research,* 17(3), 391–404. Lindeman, M., Stark, K., & Latvala, K. (2000). Vegetarianism and eating-disordered thinking. *Eating Disorders,* 8(2), 157–165. Lindeman, M. (2002). The state of mind of vegetarians: Psychological well-being or distress. *Ecology of Food and Nutrition,* 41, 75–86. Bas, M., Karabudak, E., & Kiziltan, G. (2005). Vegetarianism and eating disorders: Association between eating attitudes and other psychological factors among Turkish adolescents. *Appetite,* 44(3), 309–315. Hormes, J. M., Catanese, D., Bauer, R, & Rozin, P. (2006). Links between meat avoidance, negative eating attitudes, and disordered eating behaviors. Meeting of the American Psychological Association.

200   *Jeffrey Moussaieff Masson extols the health benefits* Jeffrey Masson has written a series of entertaining and informative books on human-animal relationships, including the bestseller *When elephants weep: The emotional lives of animals.* This quote is taken from Masson, J. M. (2009). *The face on your plate: The truth about food.* New York: W. W. Norton. (p. 168).

201   *meat-crazed society* Steiner, G. (November 22, 2009) Animal, vegetable, miserable. *New York Times.*

201   *the psychologist Jonathan Haidt* Haidt, J. (2006). *The happiness hypothesis.* New York: Basic Books. (p. 165).

202   *the no-man's land in the battle between mind and body* This line is taken from the 1997 movie *The Devil's Advocate.*

202   *the cognitive chasm between humans and chimpanzees* Marc Hauser is quoted as saying this in Balter, M. (2008). How human intelligence evolved—Is it science or 'paleofantasy'? *Science,* 319, 1028. For a contrasting perspective on the mental abilities of animals, see Hauser, M. (2000). *Wild minds: What animals really think.* New York: Henry Holt; Wynne, C. D. L. (2004) *Do animals think?* Princeton, NJ: Princeton University Press; and Bekoff, M. (2007) *The emotional lives of animals.* Noato, CA: New World Library.

205   *Carl Cohen* Cohen, C. (1987). The case for the use of animals in biomedical research. *New England Journal of Medicine,* 315, 865–870. (p. 867).

205   *Allegra Goodman* Goodman, A. (2006). *Intuition: A novel.* New York: Dial Press.

207   *you wind up killing the very creatures you have dedicated your life to studying* Note the obvious similarity in thinking between the scientists who kill animals that they devote their lives to studying and the cockfighters described in the previous chapter who are responsible for the deaths of roosters that they raise from eggs and claim to love and respect.

208   *He wrote of his pigeons* In Preece, R. (2005). *Brute souls, happy beasts, and evolution: The historical status of animals.* Vancouver, BC: University of British Columbia Press. (p. 347).

208   *"for mere damnable and detestable curiosity."* Browne, J. (2002). *Charles Darwin: The power of place.* Princeton, NJ: Princeton University Press. (p. 421).

208   *he amended the sentence* Both versions are cited in Burghardt, G. M., & Herzog, H. A. (1980). Beyond conspecifics: Is Brer Rabbit our brother? *Bioscience,* 30, 763–768.

208   *Claude Bernard, who wrote* Rudacille, D. (2000). *The scalpel and the butterfly: The war between animal research and animal protection.* New York: Farrar, Straus, and Giroux. (p. 36).

209   *Most modern ethologists agree* However, several prominent animal behaviorists have recently suggested that many similarities between the behaviors of humans and animals (e.g., tool use, the sense of fair play) are superficial and not the result of evolutionary continuity. See, for example, Bolhuis, J. J., & Wynne, C. D. L. (2009). Can evolution explain how the mind works? *Nature,* 458, 832–833.

209 *All of them said yes when it came to pain* Herzog, H. (1991). Animal conscious-ness and human conscience. *Contemporary Psychology*, 36, 7–8. The soci-ologist Mary Phillips obtained the same results in an ethnographic study of several animal research facilities. Phillips, M. T. (1993). Savages, drunks, and lab animals: The researcher's perception of pain. *Society and Animals*, 1(1), 61–81. For a perspective on the nuanced views that scientists have on the use of animal in research see Marris, E. (2006). Animal research: Grey matters. *Nature*, 444(7121), 808–810.

209 *Recent survey of 155* Knight, S., Vrij, A., Bard, K., & Brandon, D. (2009). Sci-ence versus human welfare: Understanding attitudes toward animal use. *Jour-nal of Social Issues*, 65, 463–483.

211 *the swim test was reassigned* The actual procedures used in animal experiments are often obscured by the arcane language of scientific papers. I read some of the articles on these experiments that were published in scientific journals. In no case, were the details of how the swim test was administered spelled out. For an excellent discussion of the use of language in science see Birke, L., Arluke, A., & Michael, M. (2007). *The sacrifice: How scientific experiments transform animals and people.* West Lafayette, IN: Purdue University Press.

211 *Like most scientists who use mice as models of fundamental biological pro-cesses* For an alternative opinion about the attitudes of scientists toward mice, see "World's Scientists Admit They Just Don't Like Mice," which appeared in *The Onion*, a satirical newspaper. One scientist is quoted as saying "Even just seeing them in a cage makes me feel good inside . . . I hate those little fuckers." (www.theonion.com/content/node/30800)

211 *Some by cervical dislocation* There is a macabre debate among neuroscientists about how long rodents remain conscious after decapitation. Estimates range from three seconds to over thirty seconds. This debate is discussed at length in Carbone, L. (2004). *What animals want: Expertise and advocacy in laboratory animal welfare policy.* New York: Oxford University Press.

211 *fretting over the morality of their work* The philosopher Michael Allen Fox developed a similar thought experiment using the movie *Planet of the Apes.* Fox, M. A. (1986). *The case for animal experimentation.* Berkeley: University of California Press. Fox wrote the book as a defense of animal research. How-ever, in a strange turnabout, immediately after the book was published, he changed his mind about the ethics of animal research and disavowed his own argument.

213 *the argument from marginal cases* For a succinct and accessible overview of the argument from marginal cases and other recent philosophical perspectives on the moral status of animals, see Singer, P. (2003). Animal liberation at 30. *The New York Review of Books*, 50(8), 23–26. The marginal case argument is treated at length in Dombrowski, D. A. (1997). *Babies and beasts: The argument from marginal cases.* Chicago: University of Illinois Press.

214 *Geneticists say that mouse research* See, for example, Roberts, R. B., & Thread-gill, D. W. (2005). The mouse in biomedical research. In E. J. Eisen (ed.), *The*

mouse in animal genetics and breeding research (319–140). London: Imperial College Press.

214 *the philosopher Nel Noddings* Noddings, N. (2003). *Caring: A feminine approach to ethics.* Berkeley: University of California Press. (p. 156).

214 *A 2009 Zogby survey found* The Zogby poll was conducted in February 2009 and was commissioned by the Foundation for Biomedical Research.

215 *transformation of the mouse from pest to pet to model organism* For an excellent history of the role of mice in American biomedical research see Rader, K. A. (2004). *Making mice: Standardizing animals for American biomedical research, 1900–1955,* Princeton, NJ: Princeton University Press.

215 *Founded in 1929 by Clarence Little* One of the reasons that Little settled upon mice as the ideal research animal was that he realized humans did not care about them. Critser, G. (December 2007). Of mice and men: How a twenty-gram rodent conquered the world of science. *Harper's Magazine,* 65–76.

215 *Developing a new strain can take a year and run $100,000* Waltz, E. (2005). Price of mice to plummet under NIH's new scheme. *Nature Medicine,* 11, 1261.

216 *99.5% of mouse genes have a known human counterpart* Paigen, K. (2003). One hundred years of mouse genetics: An intellectual history. II. The molecular revolution (1981–2002). *Genetics,* 163(4), 1227–1235.

216 *According to Rick Woychik* This quote is taken from a Jackson Laboratory promotional video (www.jax. org/).

217 *Carl Cohen, a University of Michigan philosopher* Cohen, C. (1987). The case for the use of animals in biomedical research. *New England Journal of Medicine,* 315, 865–870. (p. 868). For an extended defense of animal research by a prominent neuroscientist, see Morrison, A. R. (2009). *An odyssey with animals: A veterinarian's reflections on the animal rights and welfare debate.* Oxford, UK: Oxford University Press.

217 *treatments for the neglected tropical diseases* The fact is that the big pharmaceutical companies spend far less on research aimed at discovering new drugs than they do on marketing and on tweaking the molecular structure of blockbuster drugs that are on the verge of going generic. See Angel, M. (2005). *The truth about drug companies: How they deceive us and what to do about it.* New York: Random House.

217 *They say scientists have exaggerated the contributions of animal research* For an example of this point of view see Greek, R., & Greek, J. (2000). *Sacred cows and golden geese: The human cost of experiments on animals.* New York: Continuum. Barnard, N. D., & Kaufman, S. (February 1997). Animal research is wasteful and misleading. *Scientific American,* 80–82; and LaFollette, H., & Shanks, N. (1996). *Brute science: Dilemmas of animal experimentation.* London: Routledge.

218 *Researchers in Portland, Edmonton, and Albany ran eight strains of mice* This research was originally reported in Crabbe, J. C., Wahlsten, D., & Dudek, B. C.

(1999). Genetics of mouse behavior: Interactions with laboratory environment. *Science*, 284, 1670–1672. For a fascinating discussion of possible explanations of these findings, see Sapolsky, R. M. (2006). *Monkeyluv: And other essays on our lives as animals.* New York: Scribner.

218   *how much we can generalize from mice to humans* For additional articles in reputable journals questioning the rodent models, see Giles, J. (2006). Animal experiments under fire for poor design. *Nature*, 444(7122), 981; Pound, P., Ebrahim, S., Sandercock, P., Bracken, M. B., & Roberts, I. (2004). Where is the evidence that animal research benefits humans? *British Medical Journal*, 328(7438), 514–517.

218   *Writing in the journal* Immunology Davis, M. M. (2008). A prescription for human immunology. *Immunology*, 29(6), 835–838.

219   *drug that worked in four mouse studies made people with ALS sicker* Schnabel, J. (2008). Neuroscience: Standard model. *Nature*, 454(7205), 682–685.

219   *A survey conducted in England* Aldhous, P., Coghlan, A., & Copley, J. (1999). Animal experiments—where do you draw the line? Let the people speak. *New Scientist*, 162(2187), 26–31.

223   *the proliferation of surplus mice* See Critser, G. (December 2007) Of mice and men, *Harpers Magazine*, 65–76. Ormandy, E., Schuppli, C. A., & Weary, D. M. (2009). Worldwide trends in the use of animals in research: The contribution of genetically-modified animal models. *ATLA*, 37, 63–68. Rowan, A. (2007). A brief history of the animal research debate and the place of alternatives. *AATEX*, 12(3), 203–211.

223   *according to Congress, lab mice in the United States are not animals* The Animal Welfare Act excludes mice, rats, and birds that are bred specifically for use in research. Wild mice, rats, and birds, however, are covered under the AWA because they are not laboratory bred. Thus, mice and birds that are trapped in nature and brought into a lab would be covered under the AWA. However, if the wild-caught animals were used to start a breeding colony, their offspring would be exempt from coverage under the act.

224   *Judge Richey called the mouse/rat/bird exclusion* The mouse/rat/bird exclusion was originally the result of a loophole in the law that left it up to the secretary of agriculture to determine what an animal is. In 1993, Judge Ritchie ruled that the exclusion was arbitrary and capricious. However, his decision was overturned by a higher court. At the request of biotechnology interests, Senator Jesse Helms inserted last-minute language into the 2002 Farm Bill, which made the exclusion a permanent part of the Act. For a history of this legislation, see Carbone, L. (2004). *What animals want: Expertise and advocacy in laboratory animal welfare policy.* New York: Oxford University Press.

225   *exempt from the regulations* Note that birds, rats, and mice intended for research in institutions that receive funding from the National Institutes of Health are covered under the Public Health Service's *Guide for the Care and Use of Laboratory Animals*, a separate set of government animal care regula-

tions. However, mice in an estimated 800 research facilities are not covered under any federal regulations.

225   *Dogs, in contrast, are singled out for special treatment* Primates are also given special treatment. The Animal Welfare Act specifies that research facilities must look after the "psychological well-being" of monkeys and apes.

225   *Larry puts the number at well over 100 million* See Carbone, L. (2004). *What animals want: Expertise and advocacy in laboratory animal welfare policy.* New York: Oxford University Press; Taylor, K., Gordon, N., Langley, G., & Higgins, W. (2008). Estimates of worldwide laboratory animal use in 2005. *ATLA*, 36, 327–342. The estimate of 100 million mice comes from Larry Carbone who estimates that between 100 and 200 times as many mice as rats are now used in research.

227   *Our statistical analysis indicated* These results were reported in Plous, S., & Herzog, H. (2001). Reliability of protocol reviews for animal research. *Science*, 293(5530), 608–609. Responses to the article and our rejoinder can be found in *Science* (2001) 294, 1831–1832.

227   *In his novel* Zen and the Art of Motorcycle Maintenance Pirsig, R. (1974). *Zen and the art of motorcycle maintenance.* New York: William Morrow. (p. 185).

229   *imitate actions and vocalizations* Wise, S. M. (2002). *Drawing the line: Science and the case for animal rights.* Cambridge, MA: Perseus Books. (p. 157).

230   *capable of empathy* Langford, D. J., Crager, S. E., Shehzad, Z., Smith, S. B., Sotocinal, S. G., Levenstadt, J. S., et al. (2006). Social modulation of pain as evidence for empathy in mice. *Science*, 312(5782), 1967–1970.

231   *if I have calculated correctly, over 800 mice* It is difficult to actually figure out how many total animals were used in the experiments from research reports that appeared in *Science*. In addition to the published article, there were thirty additional pages of supplementary materials published online describing the details of the experiment.

232   *Marc Bekoff is an eminent ethologist* Marc made this statement in a speech to the Farm Sanctuary Hoe Down (www.farmsanctuary.org) in Orland, California, on May 16, 2009. You can see his lecture at www.youtube.com/watch?v=bZ9lq97RQ6U.

233   *the Nazi medical experiments* For discussions of the ethics of using the Nazi medical data see Moe, K. (December 1984). Should the Nazi research data be cited? *Hastings Center Report*, 5–7; and Cohen, B. (1990). The ethics of using medical data from Nazi experiments. *Journal of Halacha and Contemporary Society*, 19, 103–126.

234   *experiments on animals are also ill-gotten gains* See, for example, Regan, T. (1993). Ill gotten gains. In P. Cavalieri & P. Singer (eds.), *The great ape project.* (pp. 194–205). New York: St. Martin's Press.

237   *Joan Dunayer* Dunayer, J. (2004). *Speciesism.* Derwood, MD: Ryce Publishing. (p. 134).

237 *Jonathan Haidt* Haidt, J. (2006). *The happiness hypothesis: Finding modern truth in ancient wisdom.* New York: Basic Books.

238 *Killing animals so you can toss their bodies* Byrne's quote appears in Murphy, K. (June 13, 2009). Seattle's Pike Place fishmongers under fire. *Los Angeles Times.*

238 *The Powerful Bond between People and Pets* Anderson, P. E. (2008). *The powerful bond between people and pets: Our boundless connections to companion animals.* Westport, CT: Praeger. (p. 214).

238 *70% of animal activists* Plous, S. (1991). An attitude survey of animal rights activists. *Psychological Science,* 2(3), 194–196.

239 *an article by an ethicist named Mylan Engel* Engel, M. (2000). The immorality of eating meat. In L. P. Pojman (ed.), *The moral life: An introductory reader in ethics and literature* (pp. 856–889). New York: Oxford University Press.

240 *"non-attitudes" or "vacuous attitudes."* For more information on public opinion and attitudes toward the use of animals, see Herzog, H., Rowan, A., & Kossow, D. (2001). Social attitudes and animals. In A. N. Rowan & D. J. Salem (eds.), *The state of the animals* (pp. 55–69). Washington, DC: Humane Society Press.

240 *the majority of Americans do not get lathered up* The data in this paragraph are from a report by the Humane Research Council (March 2004), *Understanding the public image of the U. S. animal protection movement.*

241 *"psychic numbing."* Slovic, P. (2007). "If I look at the mass I will never act": Psychic numbing and genocide. *Judgment and Decision Making,* 2(2), 79–95.

241 *single red-tailed hawk from his nest* Kristof, N. (May 10, 2007). Save the Darfur puppy. *New York Times.*

241 *a series of interviews with animal activists* The results of this study of animal activists are reported in Herzog, H. A. (1993). "The movement is my life": The psychology of animal rights activism. *Journal of Social Issues,* 49, 103–119. For additional sociological and psychological research on the animal rights movement, see Herzog, H. A. (1998). Understanding animal activism. In L. Hart (ed.), *Responsible conduct with animals in research.* New York: Oxford University Press. Groves, J. M. (1997). *Hearts and minds: The controversy of laboratory animals.* Philadelphia, PA: Temple University Press. Jamison, W. V., & Lunch, W. M. (1992). Rights of animals, perceptions of science, and political activism: Profile of American animal rights activists. *Science, Technology & Human Values,* 17(4), 438–458. Plous, S. (1991). An attitude survey of animal rights activists. *Psychological Science,* 2(3), 194–196. Lowe, B. M. (2006). *Emerging moral vocabularies: The creation and establishment of new forms of moral and ethical meanings.* Oxford, UK: Lexington Books.

241 *most common thread is moral shock.* The connection between moral shock and social activism was first developed by Jasper, J. M., & Poulsen, J. D. (1995). Recruiting strangers and friends: Moral shocks and social networks in animal rights and anti-nuclear protests. *Social Problems,* 42(4), 493–512.

242 *animal rights activists are not very religious* Herzog, H. (1998). For more infor-

mation on the parallels between the animal rights movement and religion see Herzog, H. A. (1993). "The movement is my life": The psychology of animal rights activism. *Journal of Social Issues*, 49, 103–119. Jamison, W. V., Wenk, C., & Parker, J. V. (2000). Every sparrow that falls: Understanding animal rights activism as functional religion. *Society and Animals*, 8(3), 305–330. Lowe, B. M. (2006). *Emerging moral vocabularies: The creation and establishment of new forms of moral and ethical meanings.* Oxford, UK: Lexington Books.

243   *differences in people's ethical ideologies* Galvin, S. L., & Herzog, H. A. (1992). Ethical ideology, animal rights activism, and attitudes toward the treatment of animals. *Ethics & Behavior*, 2(3), 141–149.

245   *vegan-friendly condoms* Vegan-friendly condoms as well as other interesting cruelty-free "intimacy products" are available at the online Vegan Store (vegan store.com/).

248   *reserve the term "terrorist"* For the moral logic behind "direct action" from an extremist point of view see the essays in Best, S., & Nocella, A. J. (eds.). (2004). *Terrorists or freedom fighters: Reflections on the liberation of animals.* New York: Lantern Books.

248   *David Jentsch fought back* See Ringach, D. R., & Jentsch, J. D. (2009). We must face the threats. *Journal of Neuroscience*, 29, 11417–11418.

248   *the thinking of a dozen kinds of militant extremists* For discussions of the psychology of moral crusaders and terrorism, see Jasper, J. M. (1998). The emotions of protest: Affective and reactive emotions in and around social movements. *Sociological Forum*, 13(3) 397–424. Baumeister, R. F. (1999). *Evil: Inside human violence and cruelty.* New York: Henry Holt and Company, LLC. Saucier, G., Akers, L. G., Shen-Miller, S., Knezevic, G., & Stankov, L. (2009). Patterns of thinking in militant extremism. *Perspectives on Psychological Science*, 4(3), 256–271.

248   *Vlasak told an Australian television reporter* Best, S. (May 3, 2009). *Who's afraid of Jerry Vlasak?* Retrieved from civillibertarian.blogspot.com/2009/05/photo-and-caption-courtesy-of-guardian.html

249   *the FBI downplayed right-wing terrorist activities* See, for example, Dean, G. (August 31, 2006). A terrorist under every Bush? *American Chronicle.* Retrieved from www.americanchronicle.com/articles/view/13047.

250   *the species they work with* Anti-vivisectionists have concentrated their efforts on pets and primates for decades. It was an image of a dog used in research published in a 1966 *Life* magazine article that led to the passage of the first federal regulations governing animal research, and the harassment of researchers by modern animal rights activists really began in 1976 with the targeting of researcher named Lester Aronson who was conducting experiments on cats. PETA grew out of a campaign in the 1980s to close down the lab of Edward Taub, a psychologist conducting experiments on monkeys. See Morrison, A. R. (2009). *An odyssey with animals: A veterinarian's reflections on the animal rights and welfare debate.* New York: Oxford University Press.; Guillermo, K. S. (1993). *Monkey business.* Washington, DC: National Press Books.; and Blum, D. (1995). *The monkey wars.* New York: Oxford University Press.

251 *Department of Homeland Security report* Department of Homeland Security (2009) *Ecoterrorism: Environmental and animal rights militants in the United States.*

251 *the ethical theories on which the modern animal liberation movement is based* For a brief introduction to the argument for giving rights to animals see Sunstein, C. (2002). *The rights of animals: A very short primer.* John M. Olin Law & Economics Working Paper No. 157. Retrieved from papers.ssrn.com/sol3/papers.cfm? abstract_id=323661.

252 *his 1975 book* Animal Liberation If you wanted to read one book to understand the case for animal liberation, this is it. However, for a more nuanced explanation of Singer's views, see Singer, P. (1993). *Practical ethics.* Cambridge, UK: Cambridge University Press.

252 *The core of this book* Singer, P. (1975) *Animal liberation.* New York: Avon Books.

254 *legal standing for great apes* See Cavalieri, P., & Singer, P. (1993). *The great ape project: Equality beyond humanity.* New York: St. Martin's Griffin.

254 *the dog goes overboard* The lifeboat case is described in Regan, T. (1983). *The case for animal rights.* Berkeley: University of California Press. (p. 324).

254 *Joan Dunayer, author of the book* Speciesism Dunayer, J. (2004). *Speciesism.* Derwood, MD: Ryce Publishing.

255 *insects don't suffer much, he says* Singer mentioned his views on the moral status of insects in a conversation with *New York Times* columnist Nicholas Kristof. Kristof, N. (April 9, 2009). Humanity even for non-humans. *New York Times.*

255 *Singer's disrespect for chickens* Dunayer, J. (March–May 2002). Letter to the editor. *Vegan Voice.* (p. 16–17).

255 *treated equally* Singer argues that all sentient creatures deserve equal moral consideration, not equal treatment. For example, he does not believe that chimps or dogs should have the right to vote or drive.

256 *Inner Lawyer* The idea that we all have an Inner Lawyer comes from Jonathan Haidt (2007).

256 *The neuroscientist Joshua Greene* Greene, J. D., Nystrom, L. E., Engell, A. D., Darley, J. M., & Cohen, J. D. (2004). The neural bases of cognitive conflict and control in moral judgment. *Neuron, 44*(2), 389–400.

258 *"predictably irrational."* For an excellent introduction to behavioral economics see Ariely, D. (2008). *Predictably irrational: The hidden forces that shape our decisions.* New York: Harper Press.

258 *dozens of types of bias* See, for example, Lilienfeld, S. O., Ammirati, R., & Landfield, K. (2009). Giving debiasing away: Can psychological research on correcting cognitive errors promote human welfare? *Perspectives on Psychological Science, 4*(4), 390–398.

259 *In his lifeboat scenario* For a more extensive discussion of "four guys and a dog

in lifeboat" see Franklin, J. H. (2006). *Animal rights and moral philosophy.* New York: Columbia University Press.

259   *permissible to euthanize a permanently disabled infant* This argument is made on page 186 of Singer, P. (1993). *Practical ethics.* Cambridge, UK: Cambridge University Press. For a dated but fascinating profile of Singer, see Specter, M. (September 6, 1999). The dangerous philosopher. *The New Yorker.*

259   *sexual interactions between humans and animals* Singer's essay on bestiality is at Singer, P. (2001). Heavy petting. *Nerve.com.* Retrieved from www.nerve. com/Opinions/Singer/heavyPetting/. Singer was a guest on the Comedy Central television show *The Colbert Report* in December 2006. At the end of an interview that focused on the moral status of apes, Stephen Colbert dropped a question out of left field: "How about sex with animals? Is that OK with you?" Singer did not bat an eyelash. True to his beliefs that the morality of an act should be decided on the basis of whether it increases pleasure or decreases pain, Singer replied, "I'm not in favor of people having sex with animals. I think it is a lot more fun to have sex with other people."

262   *deep motivator for moral reflection and development* This quote is from an email to me from Chris Diehm.

263   *John le Carré* le Carré, J. (2008). *A most wanted man.* New York: Scribner (p. 121).

265   *Best Friends* For the story of Best Friends see Glen, S. (2001) *Best friends: The true story of the world's most beloved animal sanctuary.* New York: Kensington Books.

279   *Kwame Anthony Appiah* Appiah, K. A. (2008). *Experiments in ethics.* Cambridge, MA: Harvard University Press. (p. 199).